DATE DUE

The

Ref   lf®

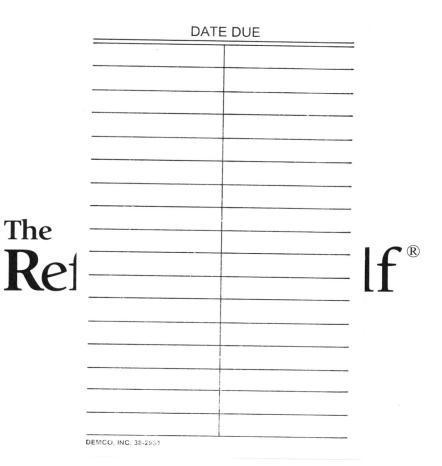

W9-CCD-685

The Next Space Age

Edited by Christopher Mari

The Reference Shelf
Volume 80 • Number 5
The H.W. Wilson Company
New York • Dublin
2008

The Reference Shelf

The books in this series contain reprints of articles, excerpts from books, addresses on current issues, and studies of social trends in the United States and other countries. There are six separately bound numbers in each volume, all of which are usually published in the same calendar year. Numbers one through five are each devoted to a single subject, providing background information and discussion from various points of view and concluding with a subject index and comprehensive bibliography that lists books, pamphlets, and abstracts of additional articles on the subject. The final number of each volume is a collection of recent speeches, and it contains a cumulative speaker index. Books in the series may be purchased individually or on subscription.

Library of Congress has cataloged this serial title as follows:

The next space age / edited by Christopher Mari.
    p. cm.—(The reference shelf ; v. 80, no. 5)
  Includes bibliographical references and index.
  ISBN 978-0-8242-1082-3 (alk. paper)
  1. Astronautics—Forecasting. 2. Outer space—Exploration. I. Mari, Christopher.
  TL790.N45 2008
  629.401'12—dc22

                                                              2008036936

Cover: This view of the rising Earth greeted the Apollo 8 astronauts as they came from behind the Moon after the lunar orbit insertion burn. The photo is displayed here in its original orientation, though it is more commonly viewed with the lunar surface at the bottom of the photo. Earth is about five degrees left of the horizon in the photo. The unnamed surface features on the left are near the eastern limb of the Moon as viewed from Earth. The lunar horizon is approximately 780 kilometers from the spacecraft. Height of the photographed area at the lunar horizon is about 175 kilometers. CREDIT: NASA

Visit H.W. Wilson's Web site: www.hwwilson.com

Printed in the United States of America

Contents

Preface   vii

I   Project Constellation: NASA's Vision for Space Exploration

Editor's Introduction   3
1. To the Moon and Beyond. Charles Dingell, William A. Johns,
   and Julie Kramer White. *Scientific American.*   5
2. To The Moon! (In a Minivan). Charles Fishman.
   Fast Company Magazine.   12
3. Can We Survive on the Moon? Guy Gugliotta. *Discover.*   19
4. Science Versus Exploration. Michael Griffin. *Ad Astra.*   26
5. Experts to Discuss U.S. Space Plan. John Schwartz.
   The New York Times.   29

II   A New Space Race? The Rise of China's Space Program

Editor's Introduction   33
1. China's Space Leadership. Leonard David. *Ad Astra.*   35
2. New Challengers Emerge, Threatening to Take the Lead.
   Guy Gugliotta. *The New York Times.*   38
3. China's Space Ambitions. Phillip C. Saunders. *Ad Astra.*   42
4. China's Space Program. William S. Murray III and
   Robert Antonellis. *Orbis.*   47
5. To Reach for the Moon. Melinda Liu and
   Mary Carmichael. *Newsweek.*   54
6. Snubbed by U.S., China Finds New Space Partners.
   Jim Yardley. *The New York Times.*   56

III   Commercializing the Cosmos: Private Industry in Outer Space

Editor's Introduction   63
1. Not Your Father's Space Program. Glenn Harlan Reynolds.
   Atlantic Monthly.   65
2. Space Travel for Fun and Profit. Katherine Mangu-Ward. *Reason.*   68
3. Space Business Booming Along the Southwest "Rocket Belt".
   Leonard David. *Ad Astra.*   77
4. Space Race II. Michael Milstein. *Smithsonian.*   83
5. Zero G, Zero Tax. Dennis Wingo. *Ad Astra.*   86

IV   Roving the Red Planet: The Continuing Exploration of Mars

Editor's Introduction   91
1. Mariner IV to Mars and More. Katherine Bracher. *Mercury.*   93
2. Inside the Mars Rover Missions. Bill Farrand. *Ad Astra.*   95

3. What We Learned on Mars. Ivan Semeniuk. *New Scientist.*   101
4. Lander Finds Ice on Mars, Scientists Say.
    David Brown. *The Washington Post.*   104
5. Life On Mars? Carl Zimmer. *Smithsonian.*   106

**V   Sailing Through the Solar System:
A Look at Unmanned Probes**
Editor's Introduction   117
1. Ask the Experts. *Scientific American.*   119
2. Spacecraft Behaving Badly. Neil deGrasse Tyson. *Natural History.*   121
3. Go Boldly, Voyager. Tim Appenzeller. *National Geographic.*   127
4. Cosmic Flock. Dan Cray. *Time.*   129

VI   Exoplanets: The Search for Extrasolar Earths
Editor's Introduction   135
1. Michel Mayor. Christopher Mari.
    Current Biography International Yearbook 2007.   137
2. Record Fifth Planet Discovered Around Distant Star.
    JR Minkel. *Scientific American.*   142
3. The New Search for Distant Planets. Geoffrey W. Marcy. *Astronomy.*   144
4. Where Is Life Hiding? Margaret Turnbull. *Astronomy.*   151
5. Searching for Earth's History Among Earth-like Worlds.
    Lisa Kaltenegger. *Mercury.*   158
6. Where Are They? Nick Bostrom. *Technology Review.*   163

Bibliography
Books   173
Web sites   176
Additional Periodical Articles with Abstracts   178

Index   185

Preface

What is a space age?

We know the first space age began on October 4, 1957 when the Soviet Union launched *Sputnik 1*, the world's first artificial satellite, into the Earth's orbit. That successful launch marked a new era of competition between the United States and the Soviet Union, in which each nation displayed its technological prowess through a series of ever more daring space firsts. After the Soviet Union launched the first man into space in April 1961, President John F. Kennedy responded by declaring that the United States would put men on the Moon by the end of the decade. Despite the Soviets' best efforts to challenge Kennedy's goal, the United States made the first manned lunar landing in July 1969 and mounted five additional landings between 1969 and 1972.

Though the competition between these Cold War adversaries produced rapid technological advances in computer design, materials science, and rocketry that quickly found applications in other fields, it did not—despite the hopes of many space advocates and lovers of science fiction—produce a lasting human presence in outer space. Earthbound concerns of the economic, political, and social variety have kept humanity close to the planetary nest since the glory days of the Apollo lunar program. While both the United States and Russia have maintained their space programs in low Earth orbit, the first space age is generally acknowledged to have ended in December 1972, when Eugene Cernan, commander of Apollo 17, became the last man to walk on the surface of the Moon.

This is not to say that there haven't been significant achievements in space exploration since that time: Robotic probes sent to the planets have given humanity a sharper understanding of our neighbors in the solar system; *Mir*, the Russian space station, and its successor, the International Space Station (ISS), have demonstrated the effects of long-term weightlessness on the body; the American space shuttle program, however imperfect, has proven that a reusable space vehicle is technologically feasible; and the Hubble Space Telescope has led to breakthroughs in the field of astrophysics, such as accurately determining the universe's rate of expansion. However, since the collapse of the Soviet Union in the early 1990s and the ratcheting down of the American space program that commenced after the tragic break-up of the space shuttle *Challenger* in 1986, our ability to send people into the cosmos has seemingly mattered less than our ability to send data streams instantaneously around the Earth. Today, the implications of the information age have replaced those of the space age in the minds of politicians and the public at large. (Ironically, advanced wireless telecommunications would not have been possible without the development of communications satellites.)

It can be argued that this second age of space exploration—characterized by the use of unmanned probes across the solar system and human activity restricted to low Earth orbit—is now drawing to a close. Humanity is poised to enter a new space age, one in which human beings will return to the Moon, visit Mars, and face off in new competitions. Following the destruction of the space shuttle *Columbia* in 2003, the United States retooled its space program to focus on a return to the Moon by the end of the next decade and a manned mission to Mars sometime thereafter. That same year China put its first "taikonaut" into orbit, becoming only the third nation to accomplish human spaceflight. The Chinese now have ambitious plans to send people to the Moon, possibly to mine its abundant resources of helium-3 for use as a nuclear fusion power source. There is also a growing movement towards privately funded spaceflight, spurred on by the successful suborbital flights of Burt Rutan's manned hybrid rocket spacecraft, *SpaceShipOne*, in 2004. As of this writing, British enterpreneur Richard Branson is planning to use Rutan's design to launch a new luxury flight service, Virgin Galactic, to send paying passengers on suborbital flights.

All of these efforts have made recent headlines in the media. However, without the constant one-upmanship that shaped the space race between the United States and the Soviet Union, such headlines are infrequent at best. If space exploration generates such little public interest, one may ask why this issue is relevant. If people have become more fascinated by cyberspace than outer space, why bother discussing the next space age? The simplest answer might be that, like it or not, the next space age is coming. Despite media skepticism and public fickleness, the space industry is thriving in both the United States and abroad. Today, in offices and factories, at engineering firms and testing facilities, through the government or the private sector, thousands of people across the globe are at work on the components that will kick-start this next space age. The purpose of this book is to provide the general reader with an overview of the most promising of these efforts. As any longtime observer of space exploration will note, great plans often die on the drawing board or after a pass through the accounting office. Consequently the programs described in this book were chosen for inclusion not only for their advanced state of preparation or execution, but also because they seem to have the greatest prospects for long-term success.

The book is divided into six chapters. The first section, "Project Constellation: NASA's Vision for Space Exploration," focuses on the National Aeronautics and Space Administration (NASA)'s plans to bring Americans out of low Earth orbit and back to the Moon, where the space agency hopes to establish a permanent lunar colony as part of its long-term preparations to conduct a manned exploration of Mars. In the second chapter, "A New Space Race? The Rise of China's Space Program," selections provide an overview of China's space program, including information about its Shenzhou space capsules and Long March rockets, as well as articles that debate the impact the Chinese program will have on its American counterpart. The issue of privately funded spaceflight is addressed in "Commercializing the Cosmos: Private Industry in Outer Space," which brings together articles that describe the many private space ventures that have sprung up in recent years. The fourth section, "Roving the Red Planet: The Continuing Exploration of Mars," assembles articles that provide historical context to the ongoing robotic exploration of the red planet, as well as information on what these missions have uncovered thus far. Robotic exploration is also examined in "Sailing Through the Solar System: A Look at Unmanned Probes," in which selected entries detail how probes work, recall famously successful probes like the Pioneer and Voyager series, and compare the various probes currently tooling about our solar system. Articles in the final section, "Exoplanets: The Search for Extrasolar Earths," discuss the efforts by astronomers to find a planet similar to Earth that might harbor extraterrestrial life.

In addition to thanking the authors who generously allowed the H.W. Wilson Company to reprint their articles, I would like to thank Joseph Miller and Paul McCaffrey for the opportunity to edit this book and Richard Stein for his tremendous contributions to its production. Special thanks go to my wife, Ana Maria Estela, and my friends Jeremy Brown, Christopher Dieckman, and Lynn Messina—a quartet of "space geeks" who helped me through this project. This book is dedicated with love to my daughter Juliana, in whose century we will see these miraculous events unfold. I hope she will be as inspired by this coming space age as I was by the first.

Christopher Mari
October 2008

1

Project Constellation:
NASA's Vision for Space Exploration

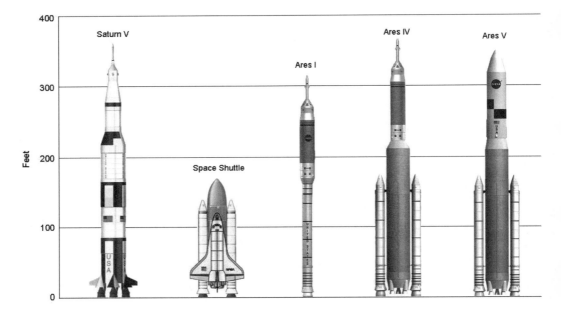

Side by side comparison of proposed Ares launch vehicles to the Saturn V, which launched Apollo astronauts to the moon, and the current space shuttle.

Editor's Introduction

An often-heard complaint about NASA's current manned space program is that it "doesn't do anything." Unlike its unmanned counterpart, which for years has produced enormous amounts of data and spectacular photographs from across our solar system and beyond, the manned program's most concrete achievement of the past decade is the construction of the International Space Station (ISS) in low Earth orbit. While the ISS is an impressive engineering feat, many observers contend that it hasn't provided much in the way of applicable scientific knowledge. Add to this a fleet of aging space shuttles and the twin tragedies associated with it, and it is not surprising to hear that many would rather use the money spent on manned spaceflight for programs that produce more tangible results.

If any event appeared to doom the U.S. manned spaceflight program, the February 2003 break-up of the space shuttle *Columbia* seemed a likely culprit. Instead of signaling the program's demise, however, the loss of a second space shuttle—coming almost 17 years to the day after the destruction of the space shuttle *Challenger*—made the American government reevaluate the priorities of its manned space program. Though no NASA personnel admitted it publicly, many had come to believe that the agency had made a mistake in the 1970s when it proposed the development of a reusable space vehicle like the shuttle, which could do little more than ferry crews and equipment into orbit. That decision, coupled with the financial considerations of a more robust presence in space, caused the American manned space program to languish in low Earth orbit since the end of the Apollo lunar missions in the early 1970s.

On January 14, 2004 President George W. Bush announced NASA's new "Vision for Space Exploration," a program, designed in response to the *Columbia* tragedy, that would free American astronauts from the confines of Earth orbit and again send them out into the solar system. The Vision's goals are straight-forward—to complete the construction of the International Space Station using the remaining three shuttles while developing a new spacecraft and launch system capable of returning astronauts to the Moon and sending them on to Mars. This new program, Project Constellation, will be the centerpiece of NASA's manned space exploration in the coming decades.

The new generation of space vehicles being built for Project Constellation will be capable of performing a variety of missions, from resupplying the ISS to landing on the Moon, possibly even traveling to a nearby asteroid. Its various components currently in development include the Ares rockets, the Orion crew capsule, the Earth Departure Stage (EDS), and the Altair lunar lander. While much of Constellation's hardware is derived from systems aboard the space shuttle, Apollo's command and service modules have influenced the design of Orion's two-part crew and service system. Additionally, the Ares rockets employ engines descended from those aboard the Saturn V, the launch vehicle that first sent astronauts to the Moon. Though designed for a return to the Moon in the latter half of the next decade, these components will also be capable of sending astronauts on a manned mission to Mars.

Project Constellation has received considerable praise from many sectors but also its fair share of criticism. Many scientists believe that the Constellation program will cut into NASA's science budget. The Mars Society questions the point of returning to the Moon and advocates going straight to Mars. Others wonder if the program, which will spread across the coming two decades and several presidential administrations, will have enough political support over that period to maintain the necessary funding. Despite these criticisms and reservations, Project Constellation is currently in development and will begin testing its components shortly after the shuttles are retired in 2010.

An overview and critiques of Project Constellation are presented in this chapter. In the first selection, "To the Moon and Beyond," Charles Dingell, William A. Johns, and Julie Kramer White offer a vivid description of how a proposed Moon mission would unfold. Charles Fishman, in his article "To the Moon! (In a Minivan)," examines how NASA is developing components of Constellation, including the Orion capsule, using a considerable amount of off-the-shelf technology. (During the heyday of Apollo, NASA had to, for example, design computers from scratch.) In an intriguing entry entitled "Can We Survive on the Moon?" Guy Gugliotta examines the challenges astronauts will face while working for extended periods on the Moon—particularly the hazards of lunar dust. In the next article, "Science Versus Exploration: A False Choice," NASA administrator Michael Griffin counters the criticism of the Vision for Space Exploration in an op-ed piece. Many of these criticisms are subsequently explored by John Schwartz in the section's final article, "Experts to Discuss U.S. Space Plan."

To the Moon and Beyond[*]

By Charles Dingell, William A. Johns, and Julie Kramer White
Scientific American Magazine, September 16, 2007

The moon, a luminous disk in the inky sky, appears suddenly above the broad crescent of Earth's horizon. The four astronauts in the *Orion* crew exploration vehicle have witnessed several such spectacular moonrises since their spacecraft reached orbit some 300 kilometers above the vast expanse of our home planet. But now, with a well-timed rocket boost, the pilot is ready to accelerate their vessel toward the distant target ahead. "Translunar injection burn in 10 seconds . . . " comes the call over the headset. "Five, four, three, two, one, mark . . . ignition. . . ."

White-hot flames erupt from a rocket nozzle far astern, and the entire ship—a stack of functional modules—vibrates as the crew starts the voyage to our nearest celestial neighbor, a still mysterious place that humans have not visited in nearly half a century. The year is 2020, and Americans are returning to the moon. This time, however, the goal is not just to come and go but to establish an outpost for a new generation of space explorers.

The Orion vehicle is a key component of the Constellation program, NASA's ambitious, multibillion-dollar effort to build a space transportation system that can not only bring humans to the moon and back but also resupply the International Space Station (ISS) and eventually place people on the planet Mars. Since the program was established in mid-2006, engineers and researchers at NASA, as well as at Lockheed Martin, Orion's prime contractor, have been working to develop the rocket launchers, crew and service modules, upper stages and landing systems necessary for the U.S. to mount a robust and affordable human spaceflight effort after its current launch workhorse, the space shuttle, retires in 2010.

To minimize development risks and costs, NASA planners based the Constellation program on many of the tried-and-true technical principles and know-how established during the Apollo program, an engineering feat that put men safely on the moon in the late 1960s and early 1970s. At the same time, NASA engineers are redesigning many systems and components using updated technology.

Orion starts with much the same general functionality as the Apollo spacecraft, and its crew capsule has a similar shape, but the resemblance is only skin-deep. Orion will, for example, accommodate larger crews than Apollo did. Four people will ride in a pressurized cabin with a volume of approximately 20 cubic meters for lunar missions (six will ride for visits to the space station starting around 2015), compared with Apollo's three astronauts (plus equipment) in a cramped volume of about 10 cubic meters.

The latest structural designs, electronics, and computing and communications technologies will help project designers expand the new spacecraft's operational flexibility beyond that of Apollo. Orion, for instance, will be able to dock with other craft automatically and to loiter in lunar orbit for six months with no one onboard. Engineers are widening safety margins as well. In the event of an emergency during launch, for example, a powerful escape rocket will quickly remove the crew from danger, a benefit space shuttle astronauts do not enjoy. But to give you a better feel for what the program involves, let us start on the ground, before the Orion crew leaves Earth. From there, we will trace the progress of a prototypical lunar mission and the technologies planned to accomplish each stage.

UP, UP AND AWAY

Towering 110 meters above the salt marshes of Florida's Kennedy Space Center, the two-stage Ares V cargo launch vehicle stands poised to blast off. The uncrewed vehicle, which contains a cluster of five powerful rocket engines, has almost the height and girth of the massive Saturn V rocket of Apollo fame. Derived from the space shuttle's external tank, Ares V's central booster tank delivers liquid-oxygen-hydrogen propellants to the vehicle's RS-68 engines—each a modified version of the ones currently used in the Delta IV military and commercial launcher. Two "strap-on," solid-fuel rocket boosters adapted from the space shuttle's system flank Ares V's central cylinder. They add the extra thrust that the launcher will need to loft the buglike lunar lander and the "Earth departure stage"—a propulsion module that contains a liquid-oxygen-hydrogen-fueled J-2X engine (a descendant of NASA's Apollo-era Saturn V J-2 motor, built by Pratt & Whitney Rocketdyne) that will enable Orion to escape Earth's gravity and travel to the moon.

Abruptly, a flash exits the tail of the Ares V, and mounds of billowing smoke clouds soon envelop the booster, gantry and launchpad. After a momentary pause, a tremendous roar echoes across the spaceport, sending birds fleeing in all directions. Slowly at first, the big rocket ascends atop an ever expanding column of gray-white exhaust. Accelerating steadily, the vehicle blazes a smoky trail across the sky and disappears into the heavens. Minutes later, amid the silence of near-Earth space, Ares V jettisons its strap-on boosters, which fall into the sea, where they will be recovered. It then sheds the protective cargo sheath that covers its nose, revealing the lunar landing module. Circling the globe at an altitude of about

300 kilometers, the robot spacecraft now awaits the next step in the lunar excursion plan: rendezvous with Orion.

That same day the four moon-bound astronauts perch 98 meters above another Kennedy launchpad, anticipating imminent liftoff. Just below their conical Orion crew capsule is a drum-shaped service module that contains the spacecraft's on-orbit propulsion engine and much of its life-support system. Protective fairings envelop both to shield them from the strong aerodynamic forces and harsh conditions they will encounter during ascent. The crew capsule and the service module sit atop NASA's two-stage Ares I crew launch vehicle. Slimmer than its big brother, the "Stick," as it is known by some, comprises another modified solid shuttle booster (constructed by Alliant Techsystems) topped by a second stage that is powered by a single J-2X motor. A spacecraft adapter serves as the structural and electrical interface between the Orion spacecraft and Ares I.

Capping the tall stack is an escape tower that is primed to rocket the occupants away from danger in the event of a failure. As the 1986 Challenger accident proved, space shuttle crews have little chance of survival if their ship sustains a major technical problem during launch and early ascent. Orion's launch-abort system (LAS), in contrast, can for a few seconds impart a thrust that is equivalent to about 15 times its own mass and that of the detached crew module. The rocket tower is set to rapidly remove the astronauts from harm's way during a mission abort while still on the launchpad or during ascent. Should a serious glitch occur on the ground, the separated system would reach an altitude of about 1,200 meters to allow for parachute deployment and a downrange, or horizontal, distance of about 1,000 meters to clear the launchpad. Mission planners estimate that the LAS, together with Orion's advanced guidance and control system, would be able to return the crew safely 999 out of 1,000 times it is needed.

But any such thoughts recede rapidly as the exhilaration of the impending launch mounts. As the countdown nears zero, commander and pilot intently eye the flight instruments on the flat-screen displays of Orion's "glass cockpit," adapted from a safety-redundant version of the avionics system used by advanced airliners such as the newly introduced Boeing 787 Dreamliner. The cockpit, with its computerized, fully electric "fly by wire" controls, energy-conserving electrical equipment and few mechanical switches, would be nearly unrecognizable to an Apollo-era astronaut.

A shudder ripples up through the entire structure, followed by a thunderous rumble. The Stick starts to move skyward. Gaining speed with every second, it rises rapidly, pressing the astronauts into their seats.

Almost two and a half minutes into the flight, the solid rocket booster is driving Ares I upward at a speed of Mach 6. At a height of about 61,000 meters, the first stage separates and falls back to Earth on parachutes so that it may be recovered and later recycled. Meanwhile the J-2X second-stage rocket motor ignites, sending the Orion crew module, the service module and the LAS through the last reaches of the atmosphere. Their usefulness ended now that the craft has exited the atmosphere, the aerodynamic shrouds break away to maximize ascent performance by shedding weight. By this time the vessel has gained enough velocity to reduce the risk of an emergency abort, so the LAS and its protective fairing also separate and fall away. The second-stage engine cuts off as the crew capsule and the service module near an altitude of about 100 kilometers.

RENDEZVOUS IN EARTH ORBIT

The service module engine then ignites, completing the job of inserting Orion into orbit and initiating the maneuvers it needs to rendezvous with the Earth departure stage and the lunar lander. Orion's main engine is adapted from the flight-proved space shuttle orbital maneuvering engine, upgraded for greater propulsion thrust and efficiency. The service module contains power generation and storage systems, radiators that expel surplus heat into space, all necessary fluids and a science equipment bay. To maximize space in the crew vehicle, the service module also carries some of the avionics system, as well as part of the environmental control and life-support subsystems. A lightweight polymer-composite honeycomb reinforced with aluminum forms its structure; simple manufacturing methods should help keep down the cost of this expendable item.

One of the more notable differences between Orion and Apollo is the addition to the service module of umbrella-shaped solar arrays that unfold when needed in orbit. Because the Apollo spacecraft was designed for moon missions measured in days, it carried hydrogen fuel cells that could generate electrical power only for relatively short periods. Orion, in contrast, must be able to produce electricity for at least six months.

Gradually, Orion catches up to the lunar lander and departure stage that Ares V had earlier placed into low Earth orbit. When the two craft finally rendezvous, the crew performs (or monitors) the final maneuvers and keeps an eye on the automated "soft capture" system as it aligns the pair and then smoothly docks them. Force-feedback and electromechanical components sense loads, automatically capturing the mating rings of the vehicles and actively damping out any contact forces. Ship and crew are now nearly ready to head for the moon.

The crew module is the only element of Orion that will make the entire trip, and it may be reused for up to 10 flights. A lightweight aluminum-lithium alloy with titanium reinforcements makes up most of the capsule structure. The exterior of the crew vehicle is lined with a thermal protection system, which, in addition to protecting its living quarters from the searing heat of reentry, also incorporates a tough, impact-resistant layer that shields it against high-velocity micrometeoroids or other debris that may strike its outer surface.

The crew module's reaction-control maneuvering system uses gaseous oxygen and methane propellants, a technology that builds on the progress engineers made during NASA's X-33 single-stage-to-orbit vehicle program, which was canceled in 2001. One advantage of the oxygen-methane propulsion system is that its fuel will be nontoxic (unlike its predecessors that used hypergolic propellants), which will help ensure the safety of the flight and ground crews after they return to Earth.

When all is ready, the Earth departure stage rocket engine ignites to propel the spacecraft toward the moon. Engineers are configuring Orion to support both "lunar sortie missions," in which crew members spend four to seven days on the moon's surface to demonstrate the Orion system's ability to transport and land humans on Earth's satellite, and "lunar outpost missions," in which a semicontinuous human presence would be established there. Because the maximum duration of a crew's stay on the lunar surface is 210 days (determined by the available supplies of oxygen, water and other consumables), Orion's continuous operation capability must exceed that period. The biggest design driver for Orion lunar missions is the amount of propellant required to meet these objectives.

After a four-day trip outbound, the crew enters into lunar orbit, having dumped the Earth departure stage along the way. The four astronauts climb into the lander, leaving the crew capsule and service module to wait for them in orbit. As with the Apollo lunar excursion module, the lunar lander consists of two components. One is the descent stage, which has legs to support the craft on the surface as well as most of the crew's consumables and scientific equipment. The other part is the ascent stage that houses the crew. After landing and exploring the surface, the foursome blasts off the moon's surface and later docks with the crew and service modules in orbit. The ascent stage of the lander is discarded into outer space, and Orion rockets back to Earth.

RETURN TO THE HOME PLANET

As the Orion astronauts close in on the blue planet, they may have to prepare for a reentry and landing quite unlike those of Apollo. Like the Gemini and Mercury spacecraft before it, Apollo splashed down in the ocean after it had plunged through the atmosphere. But because water landings would require costly fleets of recovery ships and expose a reusable spacecraft to saltwater corrosion, NASA planners may decide that Orion should touch down on land, as the Russian Soyuz spacecraft does. Orion's greater size, weight and lift, however, exacerbate the engineering challenge. The "land landing" mode is also important to minimizing life-cycle costs. If the agency instead opts to land in the ocean, Orion will be fitted with much the same capabilities as Apollo.

Unfortunately, setting down on American soil after a lunar mission presents a fundamental problem. For nearly half of the lunar month, orbital conditions would place any landing site in the Southern Hemisphere, away from the planned locations in the western continental U.S. Although the time of departure from lunar orbit can vary the longitude of the reentry point, its latitude is fixed by the declination (angular distance from the equator) of the moon relative to Earth at lunar departure. Thus, to reach landing sites in the western U.S. or waters near the continental U.S. during unfavorable periods of the lunar month, Orion will stretch its landing point into the Northern Hemisphere by employing aerodynamic lift produced as it descends into Earth's outer atmosphere. A trajectory of this type, in which a spacecraft bounces across the upper atmosphere like a stone skipping across a pond, is sometimes known as a skip reentry.

Having spent the four-day return journey from the moon fine-tuning Orion's flight path for the first crewed skip-reentry maneuver ever attempted, anticipation builds among the astronauts as the blue-white visage of our home planet grows ever larger in their view screen. They are soon occupied, however, by reorienting the ship so that the service module can be jettisoned, a necessary operation that exposes the protective heat shield on the crew module's underside. Later, after using Orion's redundant navigation system and flight computers to check that the spacecraft's attitude is positioned properly for reentry and that its trajectory is following the correct, shallow-angle route, the crew prepares for the onset of deceleration forces as Orion encounters the atmosphere.

The skip-reentry process starts out slowly. At first, the crew begins to notice weak g-forces caused by the resistance of the thin, high-altitude air. The g-forces, which push the crew members against their seats, grow steadily in strength as bits of glowing heat-shield material and streams of ionized gas streak past the windows. Shortly after Orion starts to scrape against the upper reaches of the atmosphere, the spacecraft rebounds briefly to a higher altitude. After the skip, the capsule dives deeply into the air on a path toward the landing site.

The tragic loss of the Columbia space shuttle and crew in 2003 demonstrated that the thermal protection system of a returning vehicle is critical. Atmospheric reentry generates tremendous heating on the undersurface of spacecraft (a couple of thousand degrees Celsius) caused by the friction of the air rushing by at hypersonic speeds. Because Orion's reentry velocity from a moon mission (which is on the order of 11 kilometers a second) will be 41 percent faster than a shuttle's descent speed from low Earth orbit, the heat load will be several times greater. The fact that the Orion crew module is larger than that of Apollo compounds the challenge.

The leading candidate for Orion's base heat shield is a material called PICA (phenolic impregnated carbon ablator). PICA is a matrix of carbon fibers embedded in a phenolic resin. At high temperatures, the outer surface of the PICA layer ablates, or burns away, to carry off much of that extreme heat. The ablator's surface pyrolyzes when heated, leaving a heat-resistant layer of charred material. PICA's low thermal conductivity also blocks heat transfer to the crew module. PICA was used in 2006, when it protected the Stardust spacecraft (which carried a sample from Comet Wild 2) as it came back to Earth at 13,000 meters a second—the fastest controlled reentry ever. Being 40 times larger in area, Orion's heat shield will need to be built in segments, thus adding new complexities.

LANDING ON LAND

Finally, three large parachutes—which closely resemble those used by Apollo—deploy to slow the vehicle's rate of descent. The reassuring sight of the voluminous red-and-white canopies opening above tells the astronauts that their amazing trip is almost complete. Before long, Orion is jarred by the release of its large heat shield. Hanging below the big chutes, the crew module now descends at about eight meters a second.

In the case of a "land landing," an airbag system inflates on the crew module's underside to absorb and attenuate the upcoming landing shock. With a solid jolt, the spacecraft at last sets down on dry land in the western American desert. Orion has returned home.

To The Moon!*

(In a Minivan)

By Charles Fishman
Fast Company Magazine, December 19, 2007

The essential technology America's space-shuttle astronauts depend on, which almost no one outside NASA knows about, is paper. Not just a file folder of vital checklists but actual piles of paper—stacks and stacks of it. Every minute of flight, every experiment, every space walk, is scripted. The routines are rehearsed in advance, manuals in laps, over and over. The loose-leaf sheets—called FDFs, or flight data files—are organized into functional sets, held together with three metal rings.

When the day comes to pull on the orange go-to-space suits, the paper goes too—250 pounds of it. Astronauts, strapped in for launch, have critical FDFs Velcroed to their legs for easy access. When you're hurling a 30-year-old spaceship into orbit, some things are not going to feel particularly space-age; hauling along your stacks of paper is definitely one of them.

The United States is long overdue for a new spaceship. The last time NASA's engineers sat down to design one—the space shuttle—it was 1974, and George W. Bush hadn't yet received his MBA from Harvard, or met Laura; the IBM Selectric was the dream office machine; a microwave oven was found in just 4% of U.S. kitchens.

Almost everything that matters in the world of technology and flight has changed since then: computing power, materials science, electronics, communications. Imagine if you hadn't designed something as prosaic as a car since 1974—before common use of fuel injectors, air bags, cup holders, not to mention engine-control computers and onboard navigation. A new model would likely be loaded with techno-wizardry.

Yet for NASA and Lockheed Martin, the principal contractor for designing

America's next spacecraft, the goal is simplicity, not razzle-dazzle. The nation's new spaceship is called *Orion*. In shape, it looks like a big version of a 1960s-era *Apollo* craft—a cone-shaped crew capsule atop a cylindrical service module. "This is not a Ferrari, like the space shuttle," says Skip Hatfield, NASA's project manager for the capsule. "It's more like a minivan. It's more of a vehicle to go to the grocery store in."

That is, if the grocery store is on the moon. *Orion* is part of a larger program called Constellation, which is backed by a *Jimmy Neutron*-esque slogan: "To the moon, Mars, and beyond." NASA envisions *Orion* launching a new era of American space exploration, with people living on the moon as soon as the early 2020s. The Honda Odyssey minivan is not a bad metaphor for NASA's hopes for *Orion*: reliable, functional, thoughtfully designed, with more utility than glamour.

That's what the shuttle has never been, despite its ambitions. The shuttle was sold as a space truck that would handle large cargo loads and launch twice a month. Yet in the past decade, the shuttle has averaged just four flights a year. Its systems are so temperamental that taking it to orbit has turned out to be like driving to the Grand Canyon, spending a week examining the safety of your tires and engine, then turning back and driving home with just a glance over the canyon's edge. Although the shuttle's key elements have been flying for 25 years, its technology has never moved from cutting-edge to manageable. NASA has spent a generation worrying not about where we're going in space, but about handling the capricious vehicle we're flying.

Designing any new spacecraft requires relentless innovation. But sometimes the better part of innovation is not invention, but effectiveness. And therein lies the challenge for *Orion*, and for the engineers designing it. NASA and Lockheed Martin must find the discipline to produce a straightforward spaceship, with a clear mission and mature technology. And they must do it by 2015, with a total budget of only $8 billion—the equivalent of six weeks' expenses in Iraq.

But *Orion* will hardly be primitive. Those stacks of paper the astronauts depend on, for example: They're being banished. The beloved FDFs and all the procedures they outline are being built into *Orion*'s onboard computers. But the really remarkable things are the computers themselves. The shuttle's computers had to be custom-designed. *Orion*'s computers use existing Honeywell technology. They are fifth-generation aerospace avionics boxes, with millions of hours of real-world experience, the same computers pilots use to fly Boeing 777s, hardened against vibration and radiation for the rigors of space flight.

Says Larry Price, a Lockheed engineer who is second-in-command of creating *Orion*: "We spent nothing to develop them." He's smiling. How sweet it is in the year 2007 to be designing a new rocket ship, look around, and buy the computers to fly to the moon off-the-shelf.

Building 9 of Houston's Johnson Space Center is a vast training facility with a ceiling three stories up and a floor crowded with full-size spacecraft mock-ups—the shuttle, the International Space Station. Everywhere people are using the mock-ups to train for future missions. In one corner, an astronaut in a prototype moon/Mars space suit is doing an endurance test, carting wheelbarrows of rocks up an incline, monitored by a half-dozen attendants. Tucked between space-station modules sits a squat white cone not much larger than a medium-size family camping tent and made mostly of plywood and plastic: This is the full-size *Orion* crew-capsule mock-up.

Duck through the hatch and have a seat in the capsule, and the functional austerity of *Orion* becomes vivid. It is designed to carry six people to the space station, or four to the moon. With six metal seat frames bolted in place, there is no open floor space. The capsule feels snug with four people inside; none of us are wearing space suits.

A spot has been carved out for the toilet, tucked to one side, just below floor level. For privacy, it will have a wraparound curtain. It's definitely a step up from *Apollo*—which relied on adhesive plastic bags—but, really, no more private than the third row of a minivan.

During the 1960s, NASA commanded an army of 400,000 people who were furiously designing and building *Apollo*—three times the number of Americans deployed in Iraq. Today, at Lockheed Martin, there are 1,600 people working on *Orion*, supported by another 600 at NASA. Overall, Constellation uses fewer than 5% of the number of people *Apollo* did.

Bill Johns is a senior manager for Lockheed, which won the $8 billion contract to build *Orion* in August 2006, over a joint team from Northrup Grumman and Boeing. Johns is chief engineer for the crew capsule. On the whiteboard in his office, there is only one thing boxed off with a note that says "DO NOT ERASE." In the box, in green marker, is a question: "WHAT DID *APOLLO* DO?"

The question is central to *Orion*'s unusual design philosophy. For every challenge facing *Orion*'s engineers, there is a simple mantra: Borrow or buy before you invent.

That is, borrow technology NASA has already used, if it works. Buy technology from the commercial world that has been introduced in the past three decades, technology NASA didn't have to pay to develop or debug. And if you can't find a solution in stock or off the shelf, only then do you go into the NASA workshop and mix up something new. Everything is ultimately adapted for *Orion*, but the resourcefulness provides two things the manned space program needs: efficiency and confidence.

"Does paying attention to *Apollo* limit our thinking?" Johns asks. "Yes, it does. But I don't have any lack of young engineers coming up with great new solutions to problems—I get eight or nine of those for every problem. Engineers love to reinvent things. I use that question to make sure the engineers have actually checked to see what [their predecessors'] solution was."

Examples of NASA borrowing from its own heritage are everywhere. *Orion* will use a hatch design nearly identical to *Apollo*'s. It will be launched on a solid rocket adapted from the shuttle.

Orion's parachute system, too, will be almost identical to *Apollo*'s. "I've read all the reports I can find from that time," says Koki Machin, who leads the *Orion* parachute group at NASA. "They wrote reports out the wazoo. They did a really good job with parachutes." Machin's changes are small: The heavier *Orion* will use newer material and a larger diameter, and the chutes will be tethered to *Orion* with something light, strong, and pliable like Kevlar, instead of recalcitrant braided steel cable. *Orion* is even buying the parachutes from the descendant company that made *Apollo*'s. They were a cutting-edge technology for *Apollo*, developed by a "parachute branch" that employed dozens of people. Machin's team consists of five, including him.

When *Orion*'s engineers tackle a particular design problem, they typically do something called a "trade study," in which they look beyond NASA's workshops, scanning the horizon for new solutions. So, for instance, there are two competing materials for *Orion*'s heat shield, which must protect the capsule and the astronauts upon reentry. The first is the original heat-shield material from *Apollo*, which is heavy and tedious to apply but can be used in a relatively thin layer. The second, developed in the past decade, is lighter and easier to handle but requires a thicker layer. In a NASA lab in Houston, engineers have spent the past two years evaluating the materials by blasting them in giant furnaces that can create temperatures of up to 5,000 degrees. They expect to make a choice this spring.

In space, *Orion* will get its power not from heavy fuel cells but from two circular solar panels that will unfurl on either side of the ship in space. The panels are round versions of commercial solar panels used routinely in communication and military satellites. As the performance of solar panels improves, new versions can be swapped onto *Orion*.

Seats for the astronauts are a surprisingly complicated problem. The seats need to be light and easily stowed once in space. They will be mounted on shock absorbers that allow them to cushion the impact when *Orion* bumps back to earth. Unlike *Apollo*, *Orion* will return to land, not water. Its final touchdown will be absorbed by huge air bags. But in the event that, say, one of the parachutes fails, the seats must help protect the crew. For advice on designing impact-absorbing seats, NASA has turned to NASCAR, of all outfits, which in the last few years has developed technology that restrains race car drivers and helps prevent serious injuries when their cars slam into track walls at high speed.

Every design project—a new Motorola cell phone, a new BMW dashboard, a new Manhattan skyscraper—is a series of trade-offs: between technology and functionality, between ambition and affordability, between the desires of the people creating the object and the needs of the people using it. Spacecraft design is a particularly stark version of those trade-offs because of two unusual challenges— the stakes and the laws of physics. People's lives hang on getting *Orion*'s design right. And the laws of physics impose limits terrestrial designers rarely face. Take the issue of weight. The absolute weight of a spacecraft is set early, by the size of the rocket launching it. *Orion*—service module, capsule, escape tower—must weigh no more than 50,250 pounds. The resulting cascade of trade-offs touches almost everything. There is an ongoing wrangle, for example, about whether *Orion* will have a water heater so astronauts can make coffee each morning—a slim connection to normalcy. *Apollo* had one, the shuttle has one. Is there room in *Orion*'s "weight envelope" for a water heater? What are you willing to give up to have hot coffee during a 7- to 21-day mission?

Even something as fundamental as windows depends on your perspective. Spacecraft windows have been an issue at NASA since the days of Mercury in the early '60s. Engineers would just as soon create *Orion*'s capsule without windows. That's the strongest, most efficient way to design a spacecraft's structure and skin. The astronauts would prefer a pair of bay windows. That's the way to ensure vital visibility during launch, landing, and orbital maneuvering. Although *Orion*'s flight will typically be automated, astronauts crave a sense of "situational awareness," the ability to orient themselves spatially, physically. That is critical when things start to go wrong. As astronaut Edward Lu told *Orion*'s designers, "I'll trade food for larger windows."

Yet one square foot of spacecraft window—three panes of quartz glass—weighs more than a square foot of metal hull. Every inch of window is weight that has to be shaved somewhere else.

Blaine Brown is the Lockheed engineer in charge of designing the crew capsule. He is so passionate about aeronautical design that he went out and earned a pilot's license—so he'd have a taste of what it's like to fly—and applied to be an astronaut in the class of 2000. Brown's designers delivered an initial *Orion* capsule with four main windows, two over the control panel, two on either side of it.

Astronauts assigned to consult on *Orion*'s design didn't like the windows. They were, astronaut Lee Morin says, "like looking through a mail slot"—with no view of the horizon and unsatisfactory views for docking. The astronauts originally suggested larger windows that added 80 pounds—to a spacecraft already 5,000 pounds over its limit.

Fortunately, there was a perfect arena to play out the window debate in Houston, and it illuminates the pragmatic culture that has sprung up around the *Orion* project. Squirreled away in a corner of Building 16 at Johnson sits the ROC (reconfigurable operational cockpit), a bare-bones *Orion*-capsule simulator. It is the creation of Michael Red and Alberto Sena, two NASA engineers who have worked on shuttle simulators for years and pulled together the ROC without anyone asking for it. "We just did it," Sena says. "We're trying to provide an immersion environment to aid the design."

The ROC includes just a small slice of *Orion* interior, made of white Masonite and simple aluminum framing. An ordinary bar stool with a blue-cloth seat pulls up to the control panel. Dangling overhead is a ping-pong ball on a thread. Adjust the height of the barstool so the ball rests on the bridge of your nose, pull your barstool up to the command console, and you get an astronaut's-eye view through the windows, behind which a computer plays a launch simulation on a big screen. Astronauts and designers were able to see what each of the 20 different versions of window configurations would show at critical stages of a mission.

This simple skunkworks took the guesswork out of designing the windows. "We were able to tweak them a little bit and get a lot more performance," Brown says. Because *Orion* is double-hulled (with an inner pressure shell and an outer thermal shell), the deep frames were blocking the view to each side. The astronauts were so determined to evaluate the views, Red says, that during simulations they'd end up sticking their heads right through the window holes to look around. Eventually, the windows were repositioned, and the frames were flared along the ship's hull to open up the field of view.

Total weight increase: 27 pounds. Total cost: little more than a few trips to Home Depot.

Space is a hard, unforgiving place. It will find the flaws—in thinking, in design, in human nature.

Spacecraft design is unforgiving in another important way—politically. There are plenty of critics of the course chosen by NASA administrator Mike Griffin, who since taking over in 2005 appears to be trying to get NASA to face reality: a dangerous, unreliable shuttle that is costing billions a year to keep flying; no replacement ready to take humans into space; insufficient money for robotic space science; surging space competition from other nations (notably China and India) and the private sector.

For decades, NASA's bosses and engineers have been scorched for shuttle flaws, most the result of design choices that pushed the technological envelope. Now criticism is already mounting for Constellation's perceived lack of ambition. To Griffin, *Orion* and Constellation are the way to get back to the business of space

Orion crew module mock-up at Dryden Flight Research Lab.

exploration in a rational way. "We don't have an infinite amount of money," Griffin said in a 2005 interview, as he was tightening the program's focus and timelines. "What we have is a specific task we're trying to perform, and I'm trying to do that in the simplest, cheapest, easiest, most prudent way possible."

"Space will be explored and exploited by humans," Griffin told Congress that year. "The question is, which humans, from where, and what language will they speak? It is my goal that Americans will be always among them."

Griffin is betting that *Orion* can become the symbol of a mature space program—one in which it's the destination that matters, not the transportation. The country hardly seems aware of this critical juncture, not just in 50 years of spaceflight but in 200 years of American exploration. If *Orion* does not succeed, Americans will be left grounded, for the first time in history simply shrugging at the frontier.

Orion and Constellation can't come soon enough. NASA has said that the three space shuttles will be retired at the end of 2010; *Orion* is scheduled for its first manned flight in early 2015. So if everything goes perfectly, there will be a nearly five-year gap during which the U.S. will not be able to launch its own astronauts without outside help.

Launch Pad 39-B, at Kennedy Space Center, is one of only two equipped to launch manned rockets. Up close, Pad 39-B is testament to how brutal, even primitive, our approach to space remains. The grounds of the pad encompass 40 acres of scrub brush along the Atlantic Ocean, and much of the 40 acres is blackened by each shuttle launch. Most of the smoke you see in a shuttle launch is actually steam. Starting 17 seconds before launch, a water tower cuts loose 300,000 gallons of water onto the launch pad. The water has nothing to do with heat; it acts as a sound damper. The noise from the shuttle's engines is so powerful that without the protective deluge, the shock waves would bounce off the launch pad, ricochet up, and tear the spaceship apart as it ascends. Astronauts ride a controlled explosion to orbit.

Pad 39-B has launched its share of historic missions, including the space shuttle *Challenger*, which killed seven astronauts. The first launch from 39-B was *Apollo* 10—the formal dress rehearsal for *Apollo* 11's landing on the moon—and it sent Tom Stafford and Gene Cernan to within 50,000 feet of the moon's surface. Coming home, the crew set what remains the record for the fastest manned vehicle: 24,791 miles per hour.

Pad 39-B could eventually inaugurate an era of less momentous but equally pioneering launches. It has been pulled from shuttle service and is already being rebuilt to launch *Orion*.

Can We Survive on the Moon?[*]

By Guy Gugliotta
Discover, March 21, 2007

When Neil Armstrong took "one giant leap for mankind" onto the surface of
the moon in 1969, his booted foot sank into a layer of fine gray dust, leaving an
imprint that would become the subject of one of the most famous photographs
in history. Scientists called the dust lunar regolith, from the Greek rhegos for
"blanket" and lithos for "stone." Back then scientists regarded the regolith as
simply part of the landscape, little more than the backdrop for the planting of the
American flag.

No more. Lunar scientists have learned a lot about the moon since then. They've
found that one of the biggest challenges to lunar settlement—as vexing as new
rocketry or radiation—is how to live with regolith that covers virtually the entire
lunar surface from a depth of 7 feet to perhaps 100 feet or more. It includes everything from huge boulders to particles only a few nanometers in diameter, but
most of it is a puree created by uncountable high-speed micrometeorites that have
been crashing into the moon unimpeded by atmosphere for more than 3 billion
years. A handful of regolith consists of bits of stone, minerals, particles of glass
created by the heat from the tiny impacts, and accretions of glass, minerals, and
stone welded together.

Eons of melting, cooling, and agglomerating have transformed the glass particles in the regolith into a jagged-edged, abrasive powder that clings to anything it
touches and packs together so densely that it becomes extremely hard to work on
at any depth below four inches.

For those who would explore the moon—whether to train for exploring Mars,
to mine resources, or to install high-precision observatories—regolith is a potentially crippling liability, an all-pervasive, pernicious threat to machinery and human
tissue. After just three days of moonwalks, regolith threatened to grind the joints
of the Apollo astronauts' space suits to a halt, the same way rust crippled Doro-
thy's Tin Man. Special sample cases built to hold the Apollo moon rocks lost their

vacuum seals because of rims corrupted by dust. For a permanent lunar base, such mechanical failures could spell disaster.

Regolith can play havoc with hydraulics, freeze on-off switches, and turn ball bearings into Grape Nuts. When moondust is disturbed, small particles float about, land, and glue themselves to everything. Regolith does not brush off easily, and breathing it can cause pulmonary fibrosis, the lunar equivalent of black lung. There is nothing like it on Earth. "Here you have geological processes that tend to sort and separate," says geologist Douglas Rickman of NASA's Marshall Space Flight Center. "On the moon you have meteorite impacts that mix everything together."

But space planners also see a brighter side to the story. Forty-two percent of regolith is oxygen by weight. Extract that and it will help make breathable air, rocket fuel, and, when mixed with hydrogen, water. Heat up regolith and it will harden into pavement, bricks, ceramic, or even solar panels to provide electricity. Cloak a living area in a thick enough blanket of it and it will enable astronauts to live radiation-free. If regolith is the curse of lunar exploration, it may also prove to be a blessing.

These issues lay dormant for three decades until January 2004, when President Bush announced his "Vision for Space Exploration" and gave NASA a new mandate: Return humans to the moon by 2020 and eventually send them on to Mars. More details of this plan emerged last December at a meeting of the American Institute of Aeronautics and Astronautics in Houston. Scientists are now thinking about what is needed to make the vision a reality. While there is debate about the political will to sustain lunar exploration (see "The Future of NASA," DISCOVER, September 2006), the technical hurdles are beyond dispute. The next person to step on the moon again will be taking humanity where it has never gone before, because that person will be settling in to stay—and that will be extremely hard to do.

NASA's current plans call for a series of "precursor" robotic lunar missions to test technologies and gather information. These will begin next year, long before NASA's new Orion spaceship is ready to loft its four-astronaut crew moonward. By the time that happens, perhaps around 2018, planners hope to have resolved some key unknowns: whether there are ice deposits at one of the lunar poles, whether a space suit can be made that can survive multiple journeys across the dust-ridden landscape, and whether the human body can survive dust, lengthy stays in reduced gravity, and prolonged exposure to cosmic radiation.

The first trips will be Apollo-like sorties, brief visits to test techniques and equipment and to begin building the outpost. Eventually the base will include living quarters, a launchpad, a storage facility for fuel and supplies, and a power plant. By 2024, NASA experts expect to have enough infrastructure to support a permanent human presence with four astronauts rotating every six months, the same length of a stay as on the International Space Station.

Setting up a permanent outpost on the moon would, in many respects, be more daunting than putting an outpost on Mars. Like Earth, Mars has an atmosphere, weather, and seasons, and its gravity is one-third of Earth's. The moon has one-sixth of Earth's gravity, no atmosphere, and a merciless and unending barrage of radiation and micrometeorites. Some scientists argue that if going to Mars is the ultimate goal, there's no point in going to the moon.

But if the goal is learning about long-term stays in space, going to the moon provides excellent instruction. Space station astronauts are in low Earth orbit, only 224 miles from safety. Moon astronauts will be three days from help, and Mars astronauts will, at best, be months away—virtually alone after liftoff. The explorers will not only have to learn to live in reduced gravity in cramped spaces for prolonged periods, as in the carefully calibrated indoor environment of the space station, but they must also work outside for extended periods in potentially lethal environments they cannot control. They must make consumables like oxygen, recycle them, and recycle waste. They must be able to maintain their equipment, knowing that not only their scientific mission but their very lives may depend on their repairs. And they must be able to cope with sickness, set broken bones, perform emergency appendectomies, and, in the worst of circumstances, watch a comrade die from injury or blood loss, knowing that he or she could easily have survived with timely treatment at a terrestrial hospital.

Coping with these challenges will require an attitude adjustment and a lot of practice, and screwups are better handled closer to home. Former astronaut and U.S. senator Harrison Schmitt, the last man to walk on the moon, told delegates at a NASA-sponsored moon conference last year that humanity needed to "redevelop a deep space operational structure and discipline." Others describe the situation more bluntly. NASA, grown skittish because of the losses of space shuttles Challenger and Columbia, has become too risk-averse.

"There are things we have to decide," says University of Tennessee geochemist Lawrence Taylor, a leading moon scientist. "There's going to be a hazard, and if we think it's dangerous to go to the moon, what about Mars? You just can't bail out and go home."

The abrasive regolith is just one aspect of the moon's harsh environment. The equator promises relatively happy landings on relatively smooth surfaces, but it also guarantees temperatures that exceed 250 degrees Fahrenheit during the day and plummet below −240°F during the night—and both day and night last 14 Earth days. The Apollo astronauts did most of what they did during the lunar equivalent of early morning and forenoon—light enough to see but not as hot.

Climate is the main reason NASA announced last December that it would build its outpost near one of the lunar poles. The current favorite spot is the edge of Shackleton Crater at the moon's south pole, which is expected to feature "moderate" temperatures, between -50°F and 50°F. Shackleton also has the important advantage of being in sunlight—albeit weak sunlight—for up to 80 percent of the year. Abundant light will be crucial for generating electricity. If the base were built at the lunar equator, it would be in darkness for half of every month. During that time, solar-collecting arrays would be useless.

Another important attraction of the moon's poles is the possible presence of useful natural resources. Lunar orbiters in the 1990s detected concentrations of hydrogen, a potential resource for rocket fuel. Currently no one knows how much there is or what form it takes. Some scientists suspect that a comet may have side-swiped the moon long ago, leaving water ice buried in permanently shadowed craters. Identifying the source of the hydrogen is a key goal for the robotic missions that will precede the next landing by humans. The downside of a polar landing is that the landscape there is craggier and more forbidding than at the moon's midline, which makes landings more challenging. Nonetheless, NASA officials believe the advantages at the south pole outweigh the risk.

No matter where the base is sited, astronauts on a prolonged lunar mission must contend with low gravity and radiation. Although the muscle- and bone-weakening effects of low gravity won't be a problem during the brief initial moon missions, shielding astronauts from damaging radiation exposure will be an immediate concern.

One idea is to wrap the lunar habitat in an envelope filled with radiation-absorbing water. Another is to rig an artificial magnetic field to deflect the worst rays. The easiest solution, however, will probably be to put the regolith to work: Simply place the habitat modules in a crater and bury them under a thick layer of moon dust.

How much regolith is necessary? Nobody knows. It is conceivable that radiation will cause chain reactions below the surface of the lunar soil, producing fission products from secondary reactions that are even more harmful to human tissue than unshielded bombardment. Taylor suspects that it would take 10 feet of soil or more to insulate the astronauts.

So astronauts will have to dig into the regolith, and this will not be as easy as it sounds. First there is the challenge of getting heavy equipment into space. "We can't afford to send a 200,000-pound bulldozer to the moon," says Middle Tennessee State University civil engineer Walter Wesley Boles, a longtime student of lunar construction. "And even if we did, it would perform very poorly." Engineers will have to think small. A lunar regolith mover will be "about the size of a riding lawn mower," Boles says. NASA is holding a regolith-digging contest this May, offering a $250,000 prize to the team whose robot digs the most regolith in 30 minutes—but the excavator must weigh less than 90 pounds.

Then there are even more fundamental physics problems. Heavy machinery on Earth depends on friction and gravity to provide a stable underpinning while the machine's business end cuts, pushes, pulls, digs, scrapes, or pounds. On the moon, inertia is the same—nudge something and it will move with the same vector it has on Earth—but gravity is different. Jab too hard and the machine will jump. Twist too hard and the machine tips over.

One solution is to build a bin on the back of the bulldozer and fill it with regolith to make a counterweight before serious digging begins. Another is to outfit the bulldozer with augers, so it can screw itself into the lunar surface. Boles suggests getting rid of the blade altogether and mounting a brush or a construction sweeper that would use less force and skim the regolith one thin layer at a time.

As they excavate the moon, astronauts can count on being enveloped in clouds of dust, especially if they use a sweeper. The effects of man-made regolith dust storms on tools and equipment have been known since the backwash from Apollo 12's engines sandblasted the derelict old Surveyor 3 spacecraft lying nearby. "They found moondust in every nook and cranny," says William Larson of the Kennedy Space Center, a lead scientist and program manager in NASA's efforts to develop techniques for using lunar resources. Every artist's rendering of an imagined lunar outpost features regolith mounds that would screen vital equipment and habitat from rocket-induced dust clouds on the launchpad.

Moondust is also a major unresolved issue for NASA's next-generation space suit. During the Apollo missions, three days of abbreviated moonwalks was about the limit before zippers balked, joints stiffened, and connectors began to clog. The new astronaut explorers must have a solution that will enable them to work there. Johnson Space Center space suit engineer Amy Ross says: "We're going to have to maintain ball bearings [in the joints] and replace seals. We can't have zero tolerance, but we don't want to suck up all the astronauts' free time doing maintenance."

Space engineers are still debating whether to have astronauts don overalls for dirty work or to build a "dust porch" where astronauts can clean up before entering their living quarters. They are also grappling with how to make a suit that will not easily cut or abrade yet will weigh no more than 200 pounds on Earth—33 pounds on the moon. "It's fairly challenging," Ross acknowledges.

Despite all its hazards, regolith may hold the answer, not just for blocking out radiation but also for providing building material for a self-sustaining outpost on the moon. The key lies in particles of glass and metallic iron in the lunar soil. In the 1990s the University of Tennessee's Lawrence Taylor showed that finer samples of regolith contain enough of this material to make it useful. "One night I go downstairs and stick some of it [the regolith] in the microwave," he recalls. "I had no reason to do it. It had been tried years ago and never worked. This time it just went zap!?"

Taylor found he could melt a pile of lunar soil in 10 to 20 seconds. Then he focused a single magnetron on another sample: "With 50 watts of energy I took a one-centimeter block of lunar soil to 1700 degrees Celsius (3100°F) in 10 seconds," he says.

This result has tremendous implications. By microwaving lunar soil, astronauts could weld, or sinter, the particles together to form a serviceable foundation. If they raise the temperature, the top layers would melt and turn into a tough glass. Not only would the explorers have an instant highway, they would also mitigate the worst of the dust clouds. Regolith does not blow around by itself on the moon. Human feet or tire treads have to stir it up, and if they are traveling on pavement, the dust stops.

Taylor envisions a lunar microwave machine akin to a Zamboni that smooths the ice at a hockey game. "I can sinter the soil to a foot deep with the first set of magnetrons, then have a second set that melts the top two inches into glass," he says.

Even more important, perhaps, is a plant being built by Larry Clark of Lockheed Martin that is designed to extract oxygen from regolith. Its significance is obvious to any space engineer. Liquid oxygen makes up 75 to 80 percent of a spacecraft's fuel mass. If there is no need to bring spare oxygen from Earth, launch vehicles can be far lighter and cheaper to fly or can carry much more payload. "NASA wants us to look at making 8 metric tons [9 tons] of oxygen per year," Clark says. "That's 44 kilograms [97 pounds] per day during daylight. We could refuel two ascent vehicles per year."

Clark pondered factories in space 15 years ago and kept his ideas alive for years on a shoestring research budget. Things are different now. What he is doing in Lockheed's labs south of Denver "is not an experiment," he says. "We're taking it to the next level."

Of the many ways to make oxygen from lunar soil, Clark has chosen hydrogen reduction. It operates at relatively cool temperatures, 1300 to 1500°F. The disadvantage is that it obtains oxygen almost exclusively from iron oxides, which make up just about 10 percent of the regolith. Other, hotter processes get much higher yields. Still, Clark calculates that 100 square yards of regolith excavated to a depth of only two inches will produce 660 pounds of oxygen, enough to sustain a four-member explorer team for 75 days.

Clark's lab, with its gleaming tile floors and gentle sunlight, does not look like the moon, but his machinery is the real thing. The robot excavator is about the size of a power lawn mower, and it has steel drums with scoops mounted on them—like a steamroller with cups. When technicians punch the start button, the robot glides across the floor to a sandbox about 20 feet away. The drums lower and begin to rotate. The cups scoop up sand and feed it into a hopper on the back of the robot's platform. When the hopper is full, the robot trundles over to a "lunar lander" and dumps the sand into a plastic receptacle. Leave it alone and the robot will dig and dump all day.

In the finished product, when the excavator has filled the reservoir next to the spacecraft, an elevator will lift the soil to the reactor, which will measure only 20 inches long and be shaped like a cement mixer. There the regolith will be heated and rotated under pressure while the hydrogen percolates through it. Above 1300 degrees, the iron oxides will begin to crack, and the oxygen will combine with the hydrogen, flashing off as water vapor.

If the astronauts needed water, the process would stop at that point. If not, the vapor would enter a second chamber for electrolysis. The oxygen would be siphoned off to the lunar habitat or to fuel storage tanks, while the hydrogen would return to the reactor for reuse.

Clark hopes to test his system in a few years aboard an unmanned lunar precursor mission. He has made each piece of his factory work and is in the process of integrating the parts into a seamless whole—a bona fide oxygen plant that could largely free future moon explorers from their ties to supply ships from Earth. "Every year the mission planners come around and say, 'It's real nice, but [the entire process] has never been done before,'?" Clark says. "The next time I want to be able to say, 'Well, here it is.'"

Science Versus Exploration[*]

A False Choice

By Michael Griffin
Ad Astra, Winter 2006

Once the Vision for Space Exploration [VSE] was announced, the science community immediately said, as if with once voice, "Robotic science is exploration too!" Besides, "exploration without science is tourism!" No more "flags and footprints!" [Which is to me, by the way, a rank mischaracterization of Apollo, but I won't fight that battle here. I will note that approximately one fourth of Apollo funding was devoted to the last six scientific exploration missions to the Moon, missions that resulted in a profound increase in our understanding of the history of terrestrial planets, particularly the Earth's, and of the environment in which it and life evolved.] I'm sure you've heard all of this and more. Since the science community had never previously characterized their work in terms of "exploration," many observers concluded that the theme underlying this view was, more cynically, "Don't cut our budget to pay for human spaceflight!"

Now, certainly exploration includes and enables science, for it opens and offers new capabilities to do exciting new science in new ways from new places, and about those places. What an incredible opportunity!

But, as always, there is another view, best and most tersely captured by the President's science adviser, Jack Marburger, in his March 2006 speech at the AAS Goddard Symposium. Marburger noted that the VSE is fundamentally about bringing the resources of the solar system within the economic sphere of mankind. It is not fundamentally about scientific discovery. To me, Marburger's statement is precisely right.

So a key point must be made: Exploration without science is not "tourism"; it is far more than that. It is about the expansion of human activity beyond the Earth. Exactly this point was recently noted and endorsed by no less than Stephen Hawking, a pure scientist if ever there was one. Hawking joins those, including the chairman of the NASA Advisory Council, who have long pointed out this basic truth: The history of life on Earth is the history of extinction events, and human expansion into the solar system is, in the end, fundamentally about the survival of the species. So, to me, exploration is, in and of itself, equally as noble a human endeavor as is scientific discovery.

Now, portions of the broader scientific community feel deeply disrespected—I can think of no other word—when I or anyone says or implies that "exploration" is not primarily about "science." There exists a view that the only reason we go into space is to pursue scientific discovery. To me, that is a reason, but it is certainly not the reason.

Scientists frequently tell me that they want to "be a part" of the VSE. And that is essential. But to be a part of the VSE does not mean to collect money that would otherwise go into manned spaceflight. It means rethinking planned programs of scientific activity in light of the opportunities to be made available through a newly vigorous program of human exploration. That is exactly what our NASA Advisory Council is asking the community to do with its planned Lunar Science Workshop next year.

I have said on numerous occasions—many of you have probably heard me say it—that the VSE is not about getting more money for manned spaceflight. It is obvious that such is not going to occur. Rather, the VSE is about redirecting the money that the nation has been spending on human spaceflight but to better purposes than we have been spending it. That is the key.

Similarly, participation by science in space exploration cannot be about the transfer of money into the Science Mission Directorate [SMD]. It can only be about redirecting the money being spent in existing scientific arenas, along lines which the scientific community believes to be more productive, considering the human exploration and utilization of the Moon, Mars, and near-Earth asteroids in the coming decades. It is about refocusing our thoughts on the merits and nature of future programs, given that humans will be operating in space beyond low Earth orbit.

This is the attitude that must prevail if there is to be respect from non-scientists for the contributions science can make to exploration. And it is the attitude that must prevail if scientists are to show appropriate respect for those whose primary focus is to expand the scope of the stage upon which we humans act. If mutual respect can be developed between these two groups, they can be allies rather than adversaries in the grandest endeavor I can imagine. Scientists and non-scientists alike must remember that "exploration science" is not an oxymoron.

Finally, there is the issue of control. Many members of the scientific community fully understand that the President and Congress have made decisions about the Shuttle and International Space Station programs that will not be undone. They understand that the proportion of funding at NASA that goes to SMD is at a historic high and that they should pocket their gains over the last decade and remain quiet, lest someone notice! They understand that NASA is unlikely to grow in real terms, and that therefore many projects which all of us would like to do earlier will in fact be done, later. They get all of that.

The problem is that these folks do understand these real-world limitations, and in a world with such limitations, they want to be in charge of the distribution of resources. Put bluntly, they want to exercise the inherent authority of government to decide what is being done with the money that is available for science at NASA, but without having to come to Washington, put on a NASA badge, make all the associated sacrifices and live with the consequences of their decisions. This mostly means that when you decide to do one thing, you are also deciding not to do something else that someone else would like to do, and you have to be publicly accountable for that fact.

Let us for a moment consider the situation in the abstract. The market for scientific goods and services, while dominated in the space sciences by the government, is nonetheless a market like any other. So each year the President and Congress [mostly upon the advice of scientists] determine that the pursuit of certain goals in space and Earth science is in the best interests of the United States.

Each year, the Congress approves the purchase, through NASA, of scientific goods and services to that end. As with most markets, there are more parties desiring to provide such products than can be procured, and so a variety of closely supervised competitive procurement mechanisms are employed to determine the successful suppliers of these products. Thus from a legal contractual, and managerial perspective, members of the external scientific community are suppliers to NASA, not customers.

Others with more singular and self-interested views of NASA's purpose would like to divide and conquer us. They would like to cast the argument in the terms "Science vs. Exploration." That argument is deliberate and deceptive. I don't accept it, and I urge you to reject it as well. The VSE was wise to call for the use of both robotic scientific missions and human scientific missions in the exploration of the universe. It rightly recognized the strengths of both endeavors, and it understood the symbiosis between robotic science and human exploration that will characterize our exploration campaigns.

So, this isn't about Science vs. Exploration. We will do both, and we will succeed with both. Both will contribute greatly to increased understanding of ourselves, the environment in which we live, and the solar system and universe around us. And because of the mutually reinforcing relationship between the two, we will do both better than we could do either one alone. This will be a productive partnership, and the sooner we recognize that, the better that partnership will be.

Experts to Discuss U.S. Space Plan[*]

Back to the Moon? Push on to Mars? Visit an asteroid?

By John Schwartz
The New York Times, February 12, 2008

At Stanford University on Tuesday, 50 space experts and advocates from the National Aeronautics and Space Administration, industry, academia and advocacy groups are gathering to ask whether the United States is on the right track in its plans to reach the Moon by 2020, build a long-term lunar base there and eventually send humans to Mars.

Louis Friedman, a founder of the Planetary Society, a space exploration advocacy group, said one reason for the meeting was that a new president and Congress would be coming to Washington next year. And whoever that may be, he said, "there are new political forces coming in that are not wedded to the vision for space exploration" put forth in 2004 by President Bush.

Dr. Friedman, a host of the meeting, said that with the tight budget for space exploration, it was a good time to hash out ideas to prepare for the transition.

This is not the only such effort, but it is the one that has gotten the most attention. Early coverage of the meeting, which was first described last month by the magazine Aviation Week and Space Technology, suggested that even before the meeting, those attending were ready to recommend scrapping proposals for a long-term lunar base and moving on quickly to explore asteroids.

That coverage generated controversy and startled the hosts. "You guys are describing a tempest in a teapot, and the tea isn't even brewing yet," Dr. Friedman said last month.

G. Scott Hubbard, a co-host who is a consulting professor in aeronautics and astronautics at Stanford, insisted that the slate for the meeting was clean and that it was intended to address specific proposals and the overall drain of science resources from fields like aeronautics and Earth observation science.

"We are in a double bind," he said. "As a nation we cannot abandon human exploration to other countries, nor can we further consume the remaining science budget and relinquish our roles in Earth observation, solar system exploration and space astrophysics."

To Dr. Friedman, the fact that interest was so high in a workshop that might offer alternatives to the Bush plan suggested that "latent dissatisfaction" with what has been called "Apollo on steroids" might be high as well. He questioned the Moon plan, which is taking place as other nations, including China, have announced their own lunar efforts. "What I don't understand is, we're entering a space race we've already won," he said, "and this time we might lose."

The workshop has clearly rankled Michael Griffin, NASA's administrator, who issued a response last month arguing that "the questions to be raised at this conference have been asked and answered." Many voices, he said, were heard in the planning of the program, which Congress finalized in 2005.

In an interview last week, Dr. Griffin said: "We spent three years reassessing the policy and codifying it. Changing it now? I think that's just stupid." He has suggested that some of the opposition is a sour-grapes effort by aerospace contractors who wanted a second shot at rich contracts. But, he said last week, "We don't change space policy in the United States very often—if so, you can't get anything done."

The Bush administration has ordered the shuttle program to end by 2010. It has called for a new generation of spacecraft, the Constellation program, to get humans to the Moon by 2020. But the new ships, which face enormous technical hurdles and financial squeezes, are not likely to be ready until at least 2015, and that five-year gap in American access to space could grow even bigger if the space program is sent back to the drawing board, he said.

Dr. Friedman said he was dismayed by the sharpness of NASA's reaction, since members of the workshop are "people who have worked with NASA for years, instead of against it."

One expert attending the meeting, John M. Logsdon, director of the Space Policy Institute at George Washington University, said re-examining the four-year-old vision for space exploration was "perfectly appropriate" since "this is not setting the course for the next decade, it's setting the course for the next century and beyond."

"If the plan is good, it will survive criticism," he said.

2

A New Space Race?
The Rise of China's Space Program

Shenzhou 5—*the first Chinese manned capsule.*

Editor's Introduction

Just as the rivalry between the United States and the Soviet Union shaped the early years of space exploration, it is quite possible that a similar competition between the United States and the People's Republic of China could spur on the next space age. Although China has maintained an unmanned space program since the launch of its first satellite in April 1970, it was not able to mount a serious attempt at human spaceflight until the early part of this century. Suffering from various political upheavals and isolated from the technology of both the United States and the Soviet Union during the early years of its space program, China planned and built its program in isolation and with the utmost secrecy. Typically only its successes were revealed to the media—and well after the missions had been completed.

On October 15, 2003 China became only the third nation in history to achieve manned spaceflight, when it launched taikonaut Yáng Lìwěi into orbit. The Chinese followed this accomplishment with a successful two-man mission in 2005 and are planning additional spaceflights in which taikonauts will conduct spacewalks and docking maneuvers with other vessels—two procedures that would be key to any successful Chinese lunar mission. Many observers note that while the design of China's space vehicles is based on the Russian Soyuz spacecraft, the Chinese program is mimicking many aspects of the U.S. Apollo program in order to get to the Moon.

While China cannot compete with the United States in terms of space expertise or infrastructure, the near universal popularity of it's program ensures that it will be robustly funded. Its recent successes have also earned the attention of American officials, including NASA administrator Michael Griffin. In a budget meeting on Capitol Hill, Griffin warned American legislators that the Chinese could "easily" put human beings on the Moon before Americans could return in 2018, the United States' target date. Additionally worrying to the American government was the January 2007 Chinese shootdown of an outdated weather satellite—a test likely intended to demonstrate China's missile technology and ability to threaten vulnerable U.S. military satellites. American concerns about Chinese space efforts have grown so great as to prompt a ban on selling so-called "dual-use" commercial technologies that could also have military applications to Chinese businesses.

There are many in the United States who believe that the nation has squandered over three decades of space superiority and that the Chinese are aggressively seeking to eclipse the United States in space. China's program, though seemingly slow-paced, has built incrementally on its previous achievements and is now poised to take on more ambitious projects. In addition to seeking partnerships with Russia and the European Union, China plans to assemble a space station in orbit through its Shenzhou space capsule program and launch a series of unmanned lunar probes, in preparation for manned lunar landings. An interesting aspect of the Chinese lunar program that has emerged is the country's interest in mining the Moon's regolith, the loose layer of lunar topsoil, for helium-3, a stable isotope that is a potential fuel for nuclear fusion. Though rare on Earth, an estimated one million tons of helium-3 exist on the Moon. Mining such a clean, safe, and virtually limitless energy source would give the Chinese a potent economic incentive to speed their space program and arrive on the Moon ahead of the Americans.

The articles collected in this section present an overview of the Chinese space program as well as information on the American reaction to it. In "China's Space Leadership: Decades in the Making," Leonard David provides a brief history of China's efforts in space from 1970 to 2007. Guy Gugliotta, in "New Challengers Emerge, Threatening to Take the Lead," analyzes the ways in which China's program might spur competition from the United States. In "China's Space Ambitions: Implications for U.S. Security," Phillip C. Saunders examines how China's military stands to benefit from a greater presence in space. In "China's Space Program: The Dragon Eyes the Moon (and Us)," William S. Murray III and Robert Antonellis take a closer look at the overlap between China's space program and military. Melinda Liu and Mary Carmichael discuss China's interest in mining helium-3 in "To Reach to the Moon." In the final article, "Snubbed by the U.S., China Finds New Space Partners," Jim Yardley reports on how the Chinese government has forged lucrative satellite-launching partnerships with such oil-producing countries as Venezuela and Nigeria.

China's Space Leadership[1]

Decades in the Making

By Leonard David
Ad Astra, Spring 2007

China's foothold on space began more than 35 years ago to the tune of "The East is Red." That chant for support of the political philosophy of the late Chinese leader Mao Zedong was heard from Earth orbit via Dongfanghong-1, the country's first satellite. The event placed China into a clique of countries capable of independently hurling satellites skyward, following the former Soviet Union, the United States, France and Japan.

Launched on April 24, 1970, Dongfanghong-l weighed all of 380 pounds (172 Miograms)—a satellite that still remains in Earth orbit.

Over the decades, China has orbited satellites for various purposes, built for anything from navigation needs to gathering weather information and handling telecommunications. Its burgeoning space program has also conducted launch-for-hire work for other countries.

Then there's the Shenzhou spaceship, built and flown to perform an ever-expanding agenda of human spaceflight duties—a feat that has placed China in another upper echelon of space status as the third nation independently able to launch crews into Earth orbit. Additionally, Chinese space scientists are targeting the Moon this year with Chang' e-I orbiter, a first step in a methodical program of lunar exploration.

China's military interests in space have grown over the decades too. Photo-reconnaissance from on high is but one element to further Chinese military advantage. Moreover, perhaps to underscore its own control of the space domain, China carried out a controversial antisatellite test shot in January of this year.

INCREMENTAL BUT CONSISTENT

The liftoff of Dongfanghong-1 propelled the country forward as a global space leader.

Today, China's space endeavors indicate that its program has been incremental but consistent, explains Joan Johnson-Freese, professor and chair of the Department of National Security Decision-Making at the Naval War College, in Newport, Rhode Island.

Even in the midst of the Cultural Revolution (1966–1976), when Chinese space scientists and engineers were imprisoned and sometimes killed—leading Mao to put many under state protection—the Chinese managed to become the fifth nation to launch a satellite, Johnson-Freese observes.

China's commitment to space development is clear, Johnson-Freese adds, though clearly on a different—more long-term—timetable than that of the United States. This incremental approach seems to have ultimately paid off as now the perception has been created—a perception not matching reality in this case though—that China, the third country capable of human spaceflight, is challenging the U.S. for leadership in space, she told *Ad Astra*.

PARITY WITH THE OTHER GREAT POWERS

China was one of the earliest space powers. That fact is obscured by the turmoil of China's Cultural Revolution, points out James Lewis, director and senior fellow of the Technology and Public Policy Program at the Center for Strategic and International Studies, in Washington, D.C.

The Chinese program began in the late 1950s, when Tsien Hsue-shen was deported from the United States—in exchange for U.S. prisoners of war in the Korean conflict—and began a missile program for the People's Republic of China (PROC) that produced a first launch in the early 1960s.

The 1970 blastoff of Dongfanghong-1 was followed by ambitious plans for a manned program, but this came to grief, Lewis says, due to political turmoil and because it might have been beyond the reach of China's scientific establishment.

"I think the goal all along was to show parity with the other great powers," Lewis suggests. "The Chinese talk about 'two bombs [the atomic and hydrogen bomb] and one satellite' as showing they had ended the long decline of China and were returning to their rightful place in the world. This had been a goal of Chinese nationalists since the early 1960s."

Lewis says that while China's space program has been slow going, it has been enough to achieve its primary objective, which is political.

PRIDE AT HOME, PRESTIGE ABROAD

The orbiting of Dongfanghong-1 signaled China's official entree into the Space Age—but was a byproduct of intrigue, Cold War rivalries and personalities, notes Roger Launius, chair of the Division of Space History at the Smithsonian Institution's National Air and Space Museum.

Launius advises that not an insubstantial part of China's early start in modern rocketry came through the work of Caltech's H.S. Tsien, a Chinese national. Tsien went on to become the "father" of China's missile program.

"This rocket technology made the People's Republic of China capable of entering the Space Age, and it has built on it since that time," Launius told *Ad Astra*. "In every case it has gained pride at home and prestige abroad for its successes. The capstone of that is its human program, with its first flight in 2003. It helps to define the PRC as one of the few great powers of the world, if not a superpower."

New Challengers Emerge, Threatening to Take the Lead[1]

By Guy Gugliotta
The New York Times, September 25, 2007

In March, during an otherwise routine budget hearing, Michael D. Griffin, the NASA administrator, warned members of Congress that China's aggressive space program could "easily" put humans on the Moon before American astronauts are able to return to the lunar surface under the space agency's proposed Moon-Mars project.

The China card can be a strong selling point on Capitol Hill, and Mr. Griffin, trying to finance an ambitious human spaceflight program with Mars as the ultimate goal, plays it as well as anyone.

This is America's great space-age paranoia: that the United States has frittered away 35 years of space superiority, and a new generation of rivals is about to shove it into second place.

China is the challenger du jour. It became the third nation to send a human into space in 2003, it put a two-man crew in orbit in 2005 and it plans to send an unmanned orbiter to the Moon this year. It plans to launch three astronauts and conduct its first spacewalk in 2008.

In January, China's military destroyed one of its own derelict satellites with a guided missile in a provocative demonstration of ballistic prowess. Although it is unclear what the Chinese intended, the test left no doubt that space hardware—everything from crucial navigation and communication assets to secret spy satellites—is a very soft target. And no one has more space hardware than the United States.

Despite its achievements, experts say, China is decades away from developing the full array of space expertise and infrastructure that allows the United States to simultaneously launch astronauts, send unmanned craft to explore outer planets, take spectacular pictures from orbiting telescopes and profile the eye of a hurricane. "We are leagues ahead of everyone else, and it's going to take a lot of time, effort and money to counter that superiority," said Joan Johnson-Freese, chairwoman of the Department of National Security Studies at the Naval War College.

Still, what China can do is "match, compete with or even supplant" the United States "in piecemeal ways," said Eric Hagt, head of the China program at the independent World Security Institute. By putting humans in space, China wins significant prestige, Mr. Hagt said. Putting their own man on the Moon before the United States can return would put the Chinese first in the world.

China is not the only player with space aspirations. India is developing a human spaceflight program and hopes to have astronauts in orbit by 2014. Europe and Japan, which routinely fly astronauts aboard the space shuttle, also have highly developed space science programs. Japan's innovative mission to orbit the asteroid Itokawa has produced a journal's worth of research papers, and next year the European Space Agency expects to bring its 30-satellite Galileo navigation network online as an alternative to the Global Positioning System in the United States.

Russia, once America's only rival in space, has been out of the game since the breakup of the Soviet Union. But the Russians these days are flush with petrodollars and looking for ways to reassert their power.

"Russia has the capabilities, the big industrial base, the trained cadre of people and the ability to launch in all kinds of weather," said John Logsdon, a space historian at George Washington University. "It has a relatively unambitious space plan for the next few years, but that could change."

Fifty years into the space age, human spaceflight and space science remain the most traditional measures of prestige and accomplishment, but space has also become a lucrative business and a new territory for military competition.

The big money is still in building spacecraft for governments, and the major aerospace companies worldwide, like Boeing, Lockheed, the European Aeronautic Defense and Space Company, and RKK Energia, a Russian company, focus on that. Building commercial satellites is a much more modest undertaking. Since 2004, the world has produced 80 of them; China has built 3.

Space's other cottage industry is a Russian monopoly: running a ferry service to the International Space Station with its venerable Soyuz spacecraft. Since 2001, five well-heeled space tourists have paid $20 million apiece or more to visit the station. Soyuz also carried American astronauts to the station after the 2003 Columbia accident grounded the space shuttle fleet.

With plans for the shuttle to be retired in 2010, the United States is developing new launching vehicles capable of returning astronauts to the Moon and eventually taking them to Mars. That is the kind of big, high-risk idea that has always defined space leadership, and that is where China is competing.

China launched its first satellite in 1970, but its space program languished during the chaos of the Cultural Revolution and internal power struggles. Estranged from both the Soviet Union and the United States, China learned to fend for itself. In 2003, a Long March rocket carried Lt. Col. Yang Liwei aboard the Shenzhou V spacecraft. Colonel Yang returned safely to Earth after 14 orbits and became a national hero.

The Chinese share little information about their programs, but Zhang Oinwei, who heads construction of the Shenzhou spacecraft, said the human space flight program spent about $2.3 billion in the years leading up to Colonel Yang's flight.

In 2005, Shenzhou VI, carrying two Chinese army colonels, orbited Earth for five days. Shenzhou VII, scheduled for launching in 2008, will carry three crew members who will attempt China's first spacewalks.

The Shenzhou (it means "magic vessel") spacecraft are patterned after the Soyuz but are slightly larger. China has said that its goal is to use the Shenzhou to develop orbiting skills and technology and eventually to build a small space station.

Although the Chinese have not formally announced plans to send humans to the Moon, the Shenzhou flights, coupled with plans for science missions to orbit the Moon, put robotic rovers on the lunar surface and return soil samples to Earth show that China has read the Apollo playbook.

Mr. Griffin of NASA has described the Shenzhou program as having reached the Gemini stage, recalling the two-man American flights in the mid-1960s that prepared astronauts for the Apollo missions. He told Congress in March that it would "easily be possible" for the Chinese to mount a lunar mission within a decade. If so, China could put humans on the Moon by 2018, a year earlier than the United States, under current American budget projections.

The United States, worried about China's military ambitions, has regarded the Chinese space program warily in recent years, imposing export restrictions on commercial technologies that might also have military uses.

Last year, Mr. Griffin accepted an invitation from Chinese officials to tour their space facilities in the hope of broadening ties. After cordial talks and meetings with civilian space scientists and engineers, however, Mr. Griffin cut his visit short when the military refused to let him visit the human spaceflight training facilities at the Jiuquan Satellite Launch Center in the Gobi Desert.

"If we are to conduct human space flight activity together, we have to have a great degree of trust, a great degree of sharing, a great degree of openness," Mr. Griffin told reporters at the end of his trip. "Transparency and openness mean being able to see and touch and ask questions and get answers, and China and the U.S. are not at that point."

Despite China's achievements and aspirations, there is no evidence that the Chinese can soon compete with the United States in technical skill or space infrastructure. "This is about political will," said Dr. Johnson-Freese of the Naval War College. "They are challenging what we are willing to do, or willing to pay for."

Congress seems solidly allied with Mr. Griffin on the need to finance the Moon-Mars project, but the Bush administration's insistence on paying for the program without increasing NASA's overall budget has put some spaceflight advocates on edge. At the same time, scientists interested in robotic missions and space research worry that Mr. Griffin's interest in spaceflight will shortchange those programs.

For those concerned about China, however, there was plenty to worry about after the Chinese shot down a Fengyun-1C weather satellite in January. The destruction of the satellite created more than 900 new trackable pieces of debris, increasing by 10 percent the amount of hazardous space junk circling Earth. China made matters worse by refusing for 12 days to acknowledge responsibility for the satellite's destruction, which was viewed by most of the rest of the world as pure folly.

Western analysts have offered several reasons for China's wanting to shoot down the satellite. China may have wanted to test its ability to destroy enemy satellites in a confrontation over Taiwan; it may have wanted to show the United States how vulnerable satellites are and encourage participation in talks aimed at banning weapons in space; or, it may have just been a long-scheduled test that the military conducted without informing the government.

Some considered the incident as evidence of another cold war rivalry. Representative Dana Rohrabacher, Republican of California, long a space program advocate, said the Chinese "are already in a space race with us, and we haven't recognized it."

But others say the time is ripe to sit down with the Chinese and co-opt them. Dr. Johnson-Freese said the United States should encourage China to put more satellites in space, to "make them as dependent on space assets as we are," and just as vulnerable.

China's Space Ambitions[1]

Implications for U.S. Security

By Phillip C. Saunders
Ad Astra, Spring 2005

China's October 2003 manned spaceflight highlighted its dramatic achievements in space technology. Although Chinese space technology is not state-of-the-art, China differs from other developing countries by having a space program that spans the full range of capabilities from satellite design to launch services. China builds satellites on its own, and is involved in international commercial and scientific collaborations with Europe, Russia and Brazil. The People's Republic of China has a robust commercial satellite launch industry capable of launching payloads into geosynchronous and polar orbits. Its space program is also notable for the movements of personnel and technology between the civilian and military sectors.

Beijing's space aspirations pose significant security concerns for Washington. Most of China's space programs have commercial or scientific purposes, but improved space technology could significantly improve Chinese military capabilities. China may also seek to offset U.S. military superiority by targeting U.S. space assets. This article reviews Chinese efforts to exploit space for military purposes, explores the potential for China to attack U.S. military use of space, and considers whether a Sino-American space race can be averted.

LEVERAGING SPACE FOR MILITARY OPERATIONS

China already employs space to support military operations in the areas of satellite communications, intelligence and navigation, albeit at a relatively basic level. Chinese space capabilities will improve in the coming decades, producing significant boosts in People's Liberation Army (PLA) military capabilities. The potential for Washington to restrict access to commercial satellite imagery or satellite navigation systems during a crisis is an important rationale for China to develop independent capabilities.

Secure, redundant communications are critical if the PLA is to achieve its stated objective of winning local wars under "informationalized" conditions. China employs satellites for both civilian and military communications; many satellites carry both types of signals. Satellite signals permit mobile communications and are harder to intercept or locate compared to radio communications. Commercial communications satellite programs will enhance military communications, but will not provide access to military-specific technologies such as jamming resistance and spread-spectrum transmission.

China uses satellites for the collection of photographic and electronic intelligence. China's imagery satellites use film canisters that are dropped back to earth for processing—a first-generation technology that does not provide near-real time intelligence. But the Sino-Brazilian Earth Resources Satellite program incorporates digital sensors that transmit images electronically. Low resolution limits the satellite's intelligence potential, but China is developing systems with high-resolution sensors that will provide near-real time imagery. China almost certainly exploits commercial high-resolution imagery for intelligence purposes. Chinese scientists are also exploring synthetic aperture radar technologies to provide radar imagery. China's capabilities will improve significantly as advanced technologies developed indigenously, and acquired through collaborative scientific programs, are incorporated into reconnaissance satellites.

China currently uses the U.S. global positioning system (GPS) and the Russian Glonass system, and will participate in Galileo, a European satellite navigation system. China also operates its own two-satellite Beidou system, a less sophisticated system with significant limitations for military applications. These satellites provide PLA units and weapons systems with navigation and location data that can potentially be used to improve ballistic and cruise missile accuracy and to convert "dumb bombs" into precision-guided munitions. Chinese scientists have explored using GPS signals to improve missile accuracy, but it is unclear whether current missiles employ this technology.

CHINA'S ABILITY TO DENY U.S. MILITARY USE OF SPACE

The U.S. military also makes extensive use of space for intelligence, communications, meteorology and precision targeting. Chinese analysts note that the United States employed more than 50 military-specific satellites plus numerous commercial satellites in the 2003 Iraq war. They also highlight the extensive U.S. reliance on GPS to support precision-guided munitions. The United States' space dependence will deepen as transformation and network-centric warfare increase the importance of rapid collection and dissemination of information down to tactical units and individual soldiers. Satellites also play a crucial role in U.S. missile defenses.

As U.S. dependence on space increases, concerns have grown about the potential for adversaries to attack U.S. space assets. According to current Department of Defense (DOD) doctrine, "The United States must be able to protect its space assets . . . and deny the use of space assets by its adversaries. Commanders must anticipate hostile actions that attempt to deny friendly forces access to or use of space capabilities." The 2001 Rumsfeld Commission report warned of a potential "space Pearl Harbor" if adversaries attack U.S. satellites. Underpinning these concerns is the possibility that China might target U.S. space assets in a future conflict.

Chinese strategists view U.S. dependence on space as an asymmetric vulnerability that could be exploited. As one defense analyst writes: "for countries that can never win a war with the United States by using the method of tanks and planes, attacking the U.S. space system may be an irresistible and most tempting choice." Chinese strategists have explored ways of limiting U.S. use of space, including anti-satellite [ASAT] weapons, jamming, employing lasers to blind reconnaissance satellites, and even using electro-magnetic pulses produced by a nuclear weapon to destroy satellites. A recent article highlighted Iraq's efforts to use GPS jammers to defeat U.S. precision-guided munitions.

Chinese scientists have conducted theoretical research relevant to ASAT weapons, including the use of lasers to blind satellite sensors, kinetic kill vehicles, computations for intercepting satellites in orbit, and maneuvering small satellites into close formation. Efforts to develop high-powered lasers and mobile small-satellite launch capabilities involve technologies with both commercial and ASAT applications. China probably already has sufficient tracking and space surveillance systems to identify and track most U.S. military satellites. The extent to which interest in exploiting U.S. space dependence has translated into actual ASAT development programs remains unclear. Some reports claim that Beijing is developing microsatellites or direct-ascent weapons for ASAT purposes, but the open source literature does not provide definitive proof. However, based on Chinese strategic writings, scientific research and dual-use space activities, it is logical to assume China is pursuing an ASAT capability.

IS A SINO-AMERICAN SPACE RACE AHEAD?

Efforts to exploit space for military purposes, and strategic incentives to target U.S. space assets, have put China on a collision course with a U.S. doctrine that emphasizes protecting U.S. space assets and denying the use of space by adversaries. Whether a Sino-American space race can be avoided will depend on strategic decisions by both sides and the priority placed on space control versus commercial, scientific and other military applications of space.

A key question is whether the United States can prevent potential adversaries from using space for military purposes without making its own space assets more vulnerable. United States doctrine envisions using a range of diplomatic, legal, economic and military measures to limit an adversary's access to space. However, China will almost certainly be able to use indigenous development and foreign technology to upgrade its space capabilities. Non-military means may limit Chinese access to some advanced technologies, but they will not prevent the PLA from using space.

Despite U.S. economic and technological advantages, an unrestrained space race would impose significant costs and produce few lasting strategic advantages unless the United States can dominate both offensively, by destroying an adversary's space assets, and defensively, by protecting U.S. space assets. Otherwise, the likely result would be mutual (albeit asymmetrical) deterrence, with China building just enough ASATs to threaten U.S. space capabilities. This outcome would also legitimize anti-satellite weapons.

There are some incentives to avoid confrontation. Proliferation of space weapons would inhibit scientific cooperation and raise costs of commercial satellites. (The global trend in both sectors is towards international collaboration to reduce costs.) Actual use of anti-satellite weapons could create space debris that might damage expensive commercial satellites. Commercial users of space are therefore likely to resist efforts to deploy counter-space capabilities.

Beijing's strategic incentives may also change over time. Mindful of the Soviet Union's demise due to excessive military spending, Chinese leaders are wary of entering into an open-ended space race with the United States. Moreover, as Chinese military space capabilities improve and are integrated into PLA operations, the negative impact of losing Chinese space assets may eventually outweigh the potential advantages of attacking U.S. space capabilities.

Despite incentives to avoid a space race, arms control solutions face significant obstacles. China has long advocated a treaty to prevent an arms race in outer space. The joint Sino-Russian U.N. working paper, tabled in May 2002, called for a ban on weapons in orbit and on any use of force against outer space objects. The United States has been skeptical about the utility of such a treaty, believing verification would be difficult and that it might limit future missile defense options. A ban on ASAT weapons would be one means of protecting U.S. satellites, but a verifiable ban would be hard to negotiate.

U.S. policymakers must address a number of difficult questions. Is space domination an achievable, affordable and sustainable objective? Will efforts to dissuade Beijing from developing ASAT weapons require tolerating significant improvements in Chinese military space capabilities? Can arms control protect U.S. space assets? The United States has legitimate security concerns about China's improving space capabilities, but will face tough choices in deciding on its best response.

MILESTONES IN CHINA'S QUEST FOR SPACE

1955–Tsien Hsue-Shen (Qian Xuesen), the father of modern China's space program, deported from the United States. Tsien, who came to America in 1935, helped found the U.S. space program and establish the Jet Propulsion Laboratory. In 1950, accused of being a Communist, he was placed under house arrest for almost five years. China obtained Tsien's release during Korean War prisoner exchange discussions.

1956–Tsien negotiates agreement between China and Soviet Union to exchange rocket and nuclear technology. In 1958, the Soviets deliver two of their R-2 missiles to China. The relationship ends in 1960.

1960–In September, China launches one of the Soviet R-2 missiles using domestically produced propellant. On November 5, they launch the DF-1, a homegrown variation of the R-2.

1966–China begins work on the Shuguang 1 (Dawn 1), human spaceflight program. A two-man spacecraft akin to NASA's Gemini program, its first launch was scheduled for 1973.

1968–Tsien establishes the Space Flight Medical Research Center to prepare for manned space-flight.

1969–On November 16, China launches the first in its series of domestically designed and built Long March three-stage boosters, Changzheng-1.

1970–On April 24, China launches first satellite, DFH-1. The spacecraft transmits the anthem Dong Fang Hong [The East is Red] during its 15-day mission.

1971–On March 3, China launches the Shi Jian 1 telecommunications satellite.

1974–In November, China attempts to launch the Recoverable Test Satellite, or Fanhui Shei Weixing (FSW). The launch is a failure, but between 1975 and 1987 China successfully launches nine FSW orbital missions.

1978–China announces both the Tsien Spaceplane manned program and a proposed "Skylab" type space station. Both programs are postponed in 1980 as Chinese government devotes attention and resources to economic issues.

1984–U.S. President Ronald Reagan offers to fly a Chinese astronaut aboard the space shuttle. China declines the offer.

1985–China enters the commercial space launch market. Between 1985 and 2000 they launch 27 foreign-made satellites.

1988–China successfully launches three DFH-2 military communications satellites and the Feng Yun 1A meteorological satellite. The weather satellite fails after 38 days.

1992–Tsien retires as head of China's space program, the government recommits to a manned space program and the Russian-trained Qi Faren assumes control of the agency and program. A launch date of October 1999 is scheduled for the first piloted mission.

1995–The space program suffers its most publicly known setback with the January 25 destruction of a Long March 2E that explodes 50 seconds after launch. At least six people are killed at the Xichang Satellite Launch Center, with 27 others injured by falling debris More than 80 homes were destroyed.

In March, China and Russia sign an agreement to transfer space technology. Russia agrees to help train China's astronauts.

1998–France's Aérospatiale builds Sinosat-1 for the Chinese-German company Eurospace. The satellite is launched from China in July.

1999–On November 20, China successfully launches and recovers the unmanned Shenzhou (Divine Vessel) spacecraft.

2003–On October 13, China becomes the third country to send an astronaut, Yang Liwei, into space on board Shenzhou-5.

China's Space Program[*]

The Dragon Eyes the Moon (and Us)

By William S. Murray III and Robert Antonellis
Orbis, Fall 2003

Early in 2003, before the Space Shuttle *Columbia* disaster claimed the lives of Israel's first astronaut and six American astronauts, the main headline in the world's space news for a short time was that the People's Republic of China (PRC) had successfully launched and recovered its unmanned *Shenzhou*-4 spacecraft. Beijing subsequently announced that sometime late in 2003 it would launch a manned mission into space, making it only the third country in the world, behind the United States and Russia, to do this. Its doing so would mark an important point in the PRC's progression from a third-world country to an economic and technological power.

China's emphasis on space technology has serious implications, both for the United States and for the international community in general. The PRC's space program began in the 1950s, initially serving to promote its Maoist ideology. It has transcended that ideology's decline to become a major political symbol of Chinese nationalism, an important economic sector, and an effective dual-use technology collaborator with the Chinese military. Thus the program is now more important than ever before to China's Communist regime. In 2001 Zhang Houying, human spaceflight application system commander at the Chinese Academy of Sciences, said that "by developing a human space flight program, China can aggrandize its national prestige as well as military prowess."[1]

Indeed, many of the technologies used for space exploration and utilization are similar to those necessary to place a nuclear warhead accurately on target using a ballistic missile. As the United States becomes more dependent for its own security on space assets, including a ballistic missile defense system to protect its homeland

---

[*] Reprinted from *Orbis*, Fall 2003, William S. Murray III and Robert Antonellis, "China's Space Program," pp 645–652, Copyright 2003, with permission from Elsevier.
1 Sibing He, "Space Official in Beijing Reveals Dual Purpose of Shenzhou," *Space Daily*, Mar. 7, 2003.

and allies, the PRC is increasingly turning toward utilizing its space technology to challenge the United States and to further its own military interests. "In the current and future state security strategy," a Chinese military official recently declared, "if one wants not to be controlled by others, one must have considerable space, scientific and technological strength. Otherwise one will be bullied by others."[2]

The PRC also has a record of proliferating missile technology to other countries, including North Korea and Pakistan. The Chinese space program facilitates this effort both internally, by pooling China's civilian and military know-how, and externally, by serving as a political cover for foreign technology transfers. China is a member of the Asia-Pacific Small Multi-Mission Satellite Project, which includes Iran, Pakistan, North Korea, South Korea, Thailand, Mongolia, and Bangladesh. China also has strong ties to the Russian space program, including joint projects and the development of technology and training for China's manned missions. As with the dual-use possibilities, the proliferation possibilities are numerous.

THE CHINESE SPACE PROGRAM

The world's first rockets were invented in ancient China, but the initiation of the PRC's space program required some early, albeit short-lived, technical assistance from the Soviet Union. The first modern Chinese-built spacecraft was launched in 1960. Following this launch, the Chinese space program had many ambitious plans, including putting a Chinese astronaut into space as early as 1973. However, the program suffered from the country's many political upheavals, including the Cultural Revolution. The PRC's first satellite, the DFH-1, was not launched until 1970 and did little more than transmit China's Communist anthem, "The East Is Red." In 1978 the PRC announced that it was working on a manned space capsule and a Skylab-type space station, but by 1981 these projects had been cancelled, reportedly for being too expensive.

By 1985, China's space launch technology had improved enough that it began to enter the commercial space market and, with technical assistance from the United States, developed some reliable space launch and satellite recovery capabilities. This path proved to be so successful that, by October 2000, China had developed and launched dozens of satellites, with a flight success rate of over 90 percent— making China only the fifth country or group of countries in the world (after the United States, Russia, the European Space Agency, and Japan) capable of developing and launching geo-stationary telecommunications satellites independently, and only the third country (after the United States and Russia) to utilize satellite recovery technology.[3]

The Chinese space program still lags behind its U.S. counterpart in terms of experience, expertise, and resources, but the Chinese regime is not sitting idle as the United States continues to be the dominant space power. Moving toward its goal of manned space flight, the PRC launched and recovered the unmanned *Shen-*

---

2 "China Plans First Manned Space Launch in October," *Agence Presse France*, Jan. 17, 2003.
3 "China's Space Activities," PRC Information Office of the State Council, Nov. 22, 2000.

zhou-1 (SZ-1, or *Divine Vessel-1*) on November 20-21, 1999. On January 10, 2001, the *Shenzhou-2* was launched and made 108 orbits in six days; an orbital module of the SZ-2 remained in orbit for nine months, performing a series of tests. The *Shenzhou-3* was launched on March 25, 2002, and returned a week later. With each of these launches, China increased its knowledge about and expertise in the technical challenges of putting a man in space.

The PRC has historically revealed very few details in advance of its upcoming space launches, preferring to release information only days before, if then. However, on May 18, 2002, the opening day of China's annual National Science and Technology Week,[4] Chinese officials openly discussed making the *Shenzhou-5* the country's first manned space flight. Their enthusiasm, months before the Shenzhou-4 flew, indicates that the Chinese had achieved better-than-expected progress. Prior to the SZ-3 mission, the earliest estimate Chinese officials had given for a manned mission was the SZ-6. The success of the SZ-4 prompted Chinese officials to again break with tradition and, only halfway through the SZ-4 mission, announce that it was already considered successful, and that the SZ-5 would be manned and launched in the second half of 2003. They subsequently announced that the SZ-5 launch would occur in October 2003. Until early in 2003, the names of the first Chinese astronauts were not expected to be made public until moments before the launch of the SZ-5; the Chinese news media said an earlier disclosure might "cause a psychological strain on the astronauts, which would lead to a degradation in skill."[5] However, on January 3, 2003, the *Sing Dao Times* reported that Chen Long, a military pilot, is scheduled to be one of the astronauts for the SZ-5 mission. The announcement can only be an official leak and indicates the high level of confidence the regime has in its space program.

Indeed, Chinese scientists are so confident that they have predicted a Chinese mission to the moon by 2010 and are planning missions to Mars. "We will be able to embark on a maiden unmanned mission within two and a half years, said Ouyang Ziyuan, chief scientist of the lunar program, "if the government endorses the scheme now."[6] To save time and expense, Chinese engineers are believed to be poring over the declassified archives of the U.S. Apollo program, as well as purchasing equipment and expertise from Russia. Huang Chunping, chief of the *Chang Zheng* space rocket program, has declared that "China has now solved most of the manned space technology problems, and has the capability within three to four years to step on the Moon. In fifteen years China will match the world's top level of space technology."[7]

China is also talking about building and orbiting its own space station. Official authorization to start its development was granted in February 1999 and, since then, a test chamber has been built.[8] In March 2002, Zhang Qingwei, president of the China Aerospace Science and Technology Corporation (CASC), declared that

---

4 The 2003 Science and Technology Week has been delayed until at least October and will likely be dominated by SARS-related topics.
5 Leonard David, "China's Space Program Driven By Military Ambitions," *Space.com*, Mar. 13, 2002.
6 "China Could Target Moon by 2005," *MSNBC*, Mar. 3, 2003.
7 Wei Long, "China Hopes Manned Space Flight Will Open Road to Moon," *Space Daily*, May 21, 2002.
8 "Beijing Environment, Science and Technology Update," U.S. Embassy—Beijing, Sept. 21, 2001.

China was preparing to finalize development of a new launch vehicle capable of carrying a 20-ton space station into orbit, to be launched "at an appropriate time this century."[9]

IMPLICATIONS FOR THE UNITED STATES

All of this progress in the PRC's civilian space sector has not occurred in a vacuum. China's civilian and military space programs are tightly interwoven, as is evident in the recent announcement that the SZ-4 spacecraft conducted intelligence-gathering functions during its mission and that the SZ-5 spacecraft will carry a charge-coupled device camera to be used for intelligence gathering purposes.[10] The interconnected nature of the civilian and military space programs is also seen in the structure of the CASC. Subordinate to the Chinese National Space Administration, this state-owned corporation directs five primary divisions responsible for building military missiles and civilian rockets: the Academy of Launch Vehicles (ALV), which designs and manufactures the Long March rocket series; the Academy of Space Technology, which designs and manufactures satellites; the Academy of Solid-Fuel Rockets (ASFR); the Academy of Tactical Missile Technology; and the Academy of Cruise Missile Technology.[11]

The technologies used by the ALV and the ASFR, which builds intercontinental ballistic missiles (ICBMs) targeted on the United States and other locations, are similar enough that in 1998 the U.S. Senate Select Committee on Intelligence warned that the technical assistance once provided by American companies to improve China's Long March rockets—used to launch U.S.-made satellites--may have inadvertently threatened U.S. national security by improving the accuracy and reliability of China's ICBMs.[12] The same Chinese firm that launches U.S. satellites, the China Great Wall Industry Company, has been sanctioned by the U.S. government for missile proliferation.[13] Accordingly, the United States has implemented various technology export controls and now requires that U.S. officials be present at all meetings involving technical coordination between American industry and the Chinese space launch program.

The dual-use potential of the Chinese space program is not limited to civilian rocketry and military missiles. The program's objectives by 2010 include creating an integrated military and civilian Earth observation system; building a Chinese-operated satellite broadcasting and telecommunications system, to be used for both civilian and military purposes, eventually linking the PRC's military forces;

---

9 "China Plans Heavy Lifter to Launch Space Station and More," *Space Daily*, Mar. 14, 2002.
10 Sibing He, "Space Official in Beijing Reveals Dual Purpose of Shenzhou," *Space Daily*, Mar. 7, 2003.
11 "China," *Encyclopedia Astronautica* (http://www.astronautix.com/articles/china.htm).
12 Statement of U.S. Senator Richard Shelby, Chairman, Senate Select Committee on Intelligence, June 10, 2002; "Normal Space Commercial Cooperation between China and US Must Not Be Jeopardized," Federation of American Scientists, Jan. 1, 1999 (www.fas.org/news/china/1999/990107-prc-lv.htm).
13 "China and Weapons of Mass Destruction: Implications for the United States," Conference sponsored by the U.S. National Intelligence Council (NIC) and the Federal Research Division (FRD), Library of Congress, Nov. 5, 1999.

establishing a Chinese-run Global Positioning Satellite (GPS) system; and upgrading China's Long March rocket, while continuing to develop a lower cost successor.[14]

Even as Beijing publicly declares that space should not be militarized and that space technologies should be used for peaceful purposes, military considerations play an important role in China's space program, owing in part to the program's military beginnings. For instance, China's three-stage *Chang Zheng*-1 (CZ-1) space launch vehicle (SLV) is a derivative of the military's DF-4 ballistic missile. Likewise, versions of the military's DF-5 ballistic missile have become SLVs, specifically the FB-1 in the 1970s and the contemporary *Long March* CZ-2, used to launch satellites and the *Shenzhou* series spacecraft. Among China's new ICBMs, the DF-31 has a civilian derivative for space launches, the *Kaitozhe*-1. As the *Kaitozhe* is further developed, the DF-31 will be improved, as well.

The contribution of the Chinese space program to the Chinese ICBM program is difficult to quantify. The *Long March* family of rockets includes twelve different types of launch vehicles that can deliver payloads into low-earth orbit (LEO), geo-synchronous orbit (GEO), or sun-synchronous orbit (SSO). While such capabilities are impressive, they do not necessarily translate into improved ICBM expertise. For example, satellites typically use maneuvering propellant to reach their final orbiting destinations because the SLVs that carry them are not accurate enough alone. By contrast, a nuclear warhead delivered by a ballistic missile must rely entirely upon the accuracy of that missile and upon its reentry vehicle technology. Furthermore, whereas space vehicles are generally launched in good weather conditions or else the launch is postponed, military ballistic missiles need an all-weather launch capability.

Nevertheless, the historically low accuracy of China's ballistic missiles is expected to improve as a result of advances in the Chinese space program. The estimated Circular Error Probable (CEP) of the DF-5 ICBM is more than a quarter-mile and, for the older DF-4, almost a mile. For China's theater ballistic missiles, the estimated CEP is 20–40 meters. In the near future, if not already, the use of terminal guidance systems and/or global navigation satellites could enable global positioning updates, allowing for midcourse guidance corrections. This would be especially helpful for ballistic missiles launched from mobile platforms such as the DF-31. One study done by the RAND Corporation estimated that, by using global positioning satellites (GPS), the targeting accuracy of China's ballistic missiles could be improved by 20–25 percent.[15]

The Chinese space program has become a growing factor as the United States moves to deploy a ballistic missile defense (BMD) system to protect the U.S. homeland and U.S. allies. The PRC's current nuclear deterrence doctrine emphasizes a Chinese retaliatory strike against counter-value targets (enemy cities) rather than against counterforce targets (enemy missiles that could threaten China). This is because counterforce targeting requires the use of highly accurate ballistic missiles,

---

14   Beijing Environment, Science and Technology Update, U.S. Embassy—Beijing, Nov. 30, 2001.

15   Scott Pace, et al., *The Global Positioning System: Assessing National Policies* (Santa Monica, Calif.: RAND, Critical Technologies Institute, MR-614-OSTP, 1995), p. 68.

preferably with multiple, independently targetable reentry vehicles (MIRVs)—two technologies that China currently lacks in its operational ICBMs. Without improved accuracy or MIRVed missiles, Beijing's force of only 20–30 ICBMs is too small to confidently deliver a suppressive first strike against the United States' 500 ICBMs. Once the United States deploys an effective BMD system, Beijing's minimal deterrent capability could be negated unless its force is sufficiently improved in numbers and accuracy, and by fitting its ICBMs with MIRVs. Given the impressive orbital injection accuracy that has already been achieved in China's space program, technological advances in that program may already be helping China's military to develop MIRVs. In 1981, China successfully launched, from one rocket booster, three satellites with independent orbital maneuvering. China's continued development of its capability to launch multiple payloads from a single space-launch vehicle brings it even closer to deploying MIRV technology.[16]

In the near future, the PRC also intends to field a constellation of space-based reconnaissance systems with near real-time intelligence capabilities to support its military.[17] In the event of a conflict involving, for example, Taiwan, the PRC could eventually have an advanced network in space for intelligence, targeting, early warning, and other purposes. Chinese satellites could also aid in the detection, monitoring, and targeting of U.S. military assets in East Asia and the Pacific. Improved space-based surveillance could therefore alter the military balance of power in Asia. China may already have access to advanced satellite tracking radar, as well as to anti-GPS jamming technology.[18]

The PRC's military-space strategy calls for achieving rapid access to orbital and anti-satellite weapons to seize control of the ultimate "high ground" in combat situations. China cannot yet launch a spacecraft into orbit with only 24 hours' notice, a launch-on-demand capability necessary for an effective offensive space capability. However, China's expected technological advances could give it quick access to space in the near term, as well as some ground-based and/or space-based anti-satellite (ASAT) weapons capability.[19] A Pentagon report in 1998 warned that "given China's current level of interest in laser technology, it is reasonable to assume that Beijing would develop a weapon that could destroy satellites in the future."[20] China's deployment of such an ASAT weapon could put at serious risk both the United States' space-reliant telecommunications network and the U.S. military forces that depend upon it.[21]

16 "China's Manned Spacecraft Near Ready," *Associated Press*, May 31, 2002.
17 Mark A. Stokes, *China's Strategic Modernization: Implications for the United States*, Strategic Studies Institute, Sept. 1999.
18 Annual Report on the Military Power of the People's Republic of China, Report to Congress Pursuant to the FY2000 National Defense Authorization Act, June 2000.
19 Leonard David, "China's Space Program Driven By Military Ambitions," *Space.com*, Mar. 13, 2002.
20 Bill Gertz, "Chinese Army is Building Laser Weapons," *Washington Times*, Nov. 3, 1998.
21 Paul Richter, "China May Seek Satellite Laser, Pentagon Warns," *New York Times*, Nov. 28, 1998.

A COMING ARMS RACE IN SPACE?

Beijing's stated main goal is economic development, but it continues to pursue a significant buildup and modernization of its military forces, including development of the DF-31 and DF-41 ICBMs. While the country's political leadership wants to avoid an arms race with the United States, the fact remains that the PRC aspires to become a regional hegemon and needs advanced weaponry to achieve that goal. Therefore, Chinese-manned space flights, advanced satellites, lunar exploration, and other space developments are likely to provide benefits to the PRC far beyond the expected boost in national pride and international prestige. The PRC's space objectives include a manned space mission in 2003; an increased number of improved Chinese satellites; an increased number of Chinese solid fuel rockets, perhaps eventually capable of carrying heavy payloads (e.g., 20 tons); increased missile payload capacities; MIRV technology, possibly evolving from orbital injection; improved ICBM accuracy; ground- and/or space-based ASAT weapons; an unmanned lunar mission, possibly by 2005; and a permanent space presence aboard a Chinese space station, perhaps by 2020. While the United States space program pauses in the wake of the Columbia tragedy, the Chinese space program continues to advance providing both civilian and military benefits.

The realities of international politics push for Sino-American cooperation rather than confrontation. Recent American presidents have entered office declaring a hard line against China, only to eventually develop something of a working relationship with the PRC. To effectively deal with the PRC's growing military power and its regional ambitions, the United States must craft its policies carefully, as an exclusively confrontational approach is unlikely to be sustained and a conciliatory policy will achieve undesirable results. A primary U.S. goal should be to limit technical assistance to the Chinese space program, preferably by consistently enforcing existing laws and protocols. This will not stop China's program, but it can minimize the possibility of illegal or inadvertent technology transfers and possibly retard the overall progress of China's space program.

Beijing views U.S. military power in the Pacific as an impediment to China's aspiration of becoming the dominant regional power. Beijing is modernizing and expanding China's military capabilities not only to keep an increasingly independent Taiwan in line, but also to effectively deny the U.S. military the ability to operate against China or its interests in Asia. Chinese military planners have realized that area-denial operations require the conduct of space-based surveillance and the other dual-use benefits of space technology. The Dragon is eyeing the moon because the Dragon is also eyeing us.

To Reach for the Moon[1]

By Melinda Liu and Mary Carmichael
Newsweek, October 12, 2007

Western analysts still can't say what Beijing was thinking when it shot down one of its aging weather satellites. True, the recent test was a fine show of marksmanship, destroying a refrigerator-size target sailing at orbital speed 500 miles up (as high as U.S. spy satellites). But was it worth risking a new arms race? Was it even worth the mess it caused? The Union of Concerned Scientists says the test left some 2 million pieces of shrapnel in orbit, each one a threat to any country's passing spacecraft. That's why Washington and Moscow gave up such tests decades ago: the space lanes are already littered with too much potentially lethal debris.

The drifting wreckage is a danger not only to other countries' spacecraft but to China's own ambitions for the heavens—which go far beyond blinding the U.S. military. Beijing put its first man into orbit less than four years ago. Today the Chinese are reaching for the moon. The first step, the launching of an unmanned lunar orbiter, is tentatively scheduled for April 17. A three-man mission will orbit the Earth later this year, and a spacewalk is planned for next year. Two years after that, the plan is to put down a lunar rover, followed in 2020 by a craft that will collect lunar samples and bring them home. Eventually Beijing wants to put people on the moon, although the target date remains undisclosed. "Their timetable is absolutely realistic," says Jim Benson, president of SpaceDev, a private space-exploration company in Poway, Calif. "Some of it actually seems a little conservative."

National pride is a big force behind China's moon program, but not the only one. The Chinese are aiming to do more than "just set up a flag or pick up a piece of rock," says Ye Zili of China's Space Science Society. What are they after? A limitless source of clean, safe energy to feed their voracious economy. The stable isotope helium 3 ($_3$He), a potential fuel for nuclear fusion, was first found in moon rocks brought back by the Apollo missions. It is one constituent of the "solar wind" constantly given off by the Sun. The stuff bounces off Earth's magnetic field, but the moon has no magnetic field, and its surface has been soaking up $_3$He for billions of years. If you could dig it up and put it into a fusion reactor you would get ordinary helium 4 (as in balloons), ordinary hydrogen (as in H$_2$O) and an abundance of radioactivity-free energy. According to Gerald Kulcinski, director of the Fusion Technology Institute at the University of Wisconsin at Madison, a mere 40 tons would be roughly enough to serve America's electrical needs for a year.

Or so the theory goes. The April mission is supposed to learn more about the distribution of $_3$He on the lunar surface. If significant deposits are found, China's engineers still need to design the world's first lunar mining machines and send them up—while the rest of us shrink in horror at the thought of strip mines on the moon. And meanwhile someone will have to develop a practical reactor. Kulcinski operates what he believes to be the only working model ever built, a device about the size of a basketball that sucks up far more energy than it produces. "We're not even close to breaking even," he says. What would that require? "It couldn't happen for less than tens of millions of dollars and at least 10 to 20 years," he says.

But don't count the Chinese out. "When you have a communist regime in a capitalist network, you have huge amounts of cash and the ability to direct it," says Lawrence Taylor, a director of the University of Tennessee's Planetary Geosciences Institute in Knoxville. "They could run away with this." And if they fail? At least they will have walked on the moon.

Snubbed by U.S., China Finds New Space Partners[1]

By Jim Yardley
The New York Times, May 24, 2007

For years, China has chafed at efforts by the United States to exclude it from full membership in the world's elite space club. So lately China seems to have hit on a solution: create a new club.

Beijing is trying to position itself as a space benefactor to the developing world—the same countries, in some cases, whose natural resources China covets here on earth. The latest and most prominent example came last week when China launched a communications satellite for Nigeria, a major oil producer, in a project that serves as a tidy case study of how space has become another arena where China is trying to exert its soft power.

Not only did China design, build and launch the satellite for Nigeria, but it also provided a huge loan to help pay the bill. China has also signed a satellite contract with another big oil supplier, Venezuela. It is developing an earth observation satellite system with Bangladesh, Indonesia, Iran, Mongolia, Pakistan, Peru and Thailand. And it has organized a satellite association in Asia.

"China is starting to market and sell this technology to developing countries that need it," said Shen Dingli, a professor in international relations at Fudan University in Shanghai. Of the Nigeria deal, Mr. Shen added: "It gives substance to Sino-African relations. Not only does China buy raw materials, but also we sell some things."

For China, the strategy is a blend of self-interest, broader diplomacy and, from a business standpoint, an effective way to break into the satellite market. Satellites have become status symbols and technological necessities for many countries that want an ownership stake in the digital world dominated by the West, analysts say.

"There's clearly a sense that countries like Nigeria want to have a stronger presence in space," said Peter J. Brown, a journalist who specializes in satellite technology and writes frequently about the satellite market in Asia. "As you look around the map, more and more countries are moving to get satellites up."

China's more grandiose space goals, which include building a Mars probe and, eventually, putting an astronaut on the moon, are based on an American blueprint in which space exploration enhances national prestige and advances technological development. But Beijing also is focused on competing in the $100 billion commercial satellite industry.

In recent years, China has managed to attract customers with its less expensive satellite launching services. Yet it had never demonstrated the technical expertise to compete for international contracts to build satellites.

The Nigeria deal has changed that. Chinese engineers designed and constructed the geostationary communications satellite, called the Nigcomsat-1. A state-owned aerospace company, Great Wall Industry Corporation, will monitor the satellite from a ground station in northwestern China. It will also train Nigerian engineers to operate a tracking station in Abuja, their national capital.

Last week, a day after the launching, Ahmed Rufai, the Nigerian project manager for the satellite, was exultant as he paused between appointments at his Beijing hotel. Nigeria may be rich in oil, he said, but it lacks many of the basic building blocks of a modern, information-based economy.

"We want to be part of the digital economy," Mr. Rufai said, noting that Africa suffers more than any other continent from the so-called digital divide. "We are trying to diversify the economic base of the country."

Mr. Rufai predicted that the satellite would pay for itself within seven years as Nigeria sold bandwidth to commercial users. But he also predicted major improvements for Nigeria itself: "distance learning" educational programs for remote rural areas; online public access to government records; a video monitoring system of remote oil pipelines to allow quicker responses to spills; and an online banking system.

Nigeria is a risky customer for any satellite manufacturer. It is consistently rated one of the most corrupt nations, and at least one Western aerospace company has become embroiled in business disputes there. "Business ventures with Nigeria have been difficult, to say the least," said Roger Rusch, president of TelAstra, a satellite communications consulting firm in California.

Nigeria put the project out for bidding in April 2004. Mr. Rufai said that 21 bids had arrived from major aerospace companies but that nearly all of them failed to meet a key requirement: a significant financial package.

Mr. Rufai said the Western companies saw Nigeria as a major gamble. "Their response was very cool," he said of one financial institution approached about backing the deal. "They said, 'Oh, Nigeria. Don't touch it.'"

China was not so cautious. With the satellite priced at roughly $300 million, the state-owned Export-Import Bank of China, or China ExIm, granted $200 million in preferential buyer's credits to Nigeria. The bank often provides the hard currency for China's soft power aspirations: In Africa, China ExIm has handed out more than $7 billion in loans in recent years, according to one study.

"They were the only ones who stated in concrete terms that they would be able to support the project," Mr. Rufai said. Quality remained a concern. Last year,

China suffered a major setback with the failure of the Sinosat-2. It was the most sophisticated satellite ever made in China, and it suffered a systems breakdown on its first launching. The Nigerian satellite was delayed for three months so that it could be retrofitted.

Joan Johnson-Freese, chairwoman of the Department of National Security Studies at the Naval War College, said China still trailed major aerospace companies in the quality and sophistication of its satellites, which is one reason it is marketing to developing countries. But, she added, the strategy was working on multiple levels.

"They want to play a leadership role for developing countries that want to get into space," Dr. Johnson-Freese said in an interview earlier this year. "It's just such a win-win for them. They are making political connections, it helps them with oil deals and they bring in hard currency to feed back into their own program to make them even more commercially competitive."

Satellites also are becoming vital to Beijing's domestic development plans. In the next several years, China could launch as many as 100 satellites to help deliver television to rural areas, create a digital navigational network, facilitate scientific research and improve mapping and weather monitoring. Research centers on microsatellites have opened in Beijing, Shanghai and Harbin, and a new launching center is under construction in Hainan Province.

But China's focus on satellites has also brought suspicions, particularly from the United States, since most satellites are "dual use" technologies, capable of civilian and military applications. Currently, China is overhauling its military in a modernization drive focused, in part, on developing the capacity to fight a "high tech" war.

Analysts say China's determination to develop its own equivalent to the Global Positioning System, or G.P.S., is partly because such a system would be critical for military operations if a war were to erupt over Taiwan.

Most alarmingly to Western countries, China conducted an antisatellite test in January by firing a missile into space, destroying one of its own orbiting satellites and scattering a trail of dangerous debris despite its oft-stated opposition to the use of weapons in space. Four months later, Washington is still trying to parse China's motivations, while China has offered little explanation.

Space relations between the powers were already frosty. Washington, responding to scandals over stolen technology, has tried for nearly a decade to isolate the Chinese space program through export restrictions that prohibit the use of American space technology on satellites launched in China. Washington also has prevented China from participating in the International Space Station and, in some cases, stopped Chinese scientists from attending space conferences in America.

Michael D. Griffin, NASA's administrator, did signal a thaw in relations when he visited China last fall. But critics say the American strategy has backfired. A recent critique of the Bush administration's space policy blamed Washington for alienating space allies with a "go it alone" philosophy. It also blamed the export restrictions for damaging American competitiveness and helping foreign competitors like China gain an advantage in the commercial market.

China, meanwhile, eyes the United States warily. Earlier this year, Eric Hagt, director of the China program for the World Security Institute, testified in Washington that China's increasing investment in space has made it feel more vulnerable at a time when the United States is advocating missile defense programs in the name of protecting against terrorist states.

China believes the United States is determined to dominate space, even as China's own national interests are increasingly tied to space, Mr. Hagt said. "The United States needs to come to grips with the reality that China will demand more 'strategic room' in space," he told the federal U.S.-China Economic and Security Review Commission.

The United States is also realizing that many parts of the world are happy to give China that space. When the Nigerian satellite was launched, the blastoff was televised live to Nigeria, the Chinese news media reported. Nigerian newspapers proclaimed the satellite as a seminal moment in the country's efforts to modernize its economy.

Mr. Rufai, the Nigerian project manager, said he was certain that other developing countries had noticed how China had designed, built, launched and financed the satellite.

"It's a model that people will try to replicate," he said.

3

Commercializing the Cosmos:
Private Industry in Outer Space

SpaceShipOne, *the first privately funded spacecraft, in fight.*

Editor's Introduction

While some imagine the next space age will be shaped by a potential Sino-American space race, others believe that one need only look at the flourishing world of private space ventures to glimpse the future of space exploration. During the first space age, only governments had the finances, infrastructure, and technical know-how to mount space missions. Today, many space-faring governments work in conjunction with private contractors to design and build spacecraft and to launch commercial satellites. In recent years, however, space enthusiasts and entrepreneurs have moved beyond government sponsorship, partnering with one another to design and build sub-orbital spacecraft that will make personal spaceflight possible for private citizens.

While such "space planes" have not yet become as commonplace as the ones depicted by Stanley Kubrick in his 1968 film *2001: A Space Odyssey*, they are nevertheless moving quickly from the realm of science fiction to science fact. As of this writing, the visionary airplane designer Burt Rutan is out in the Mojave Desert putting the finishing touches on a vehicle being funded by Richard Branson that will conduct regular sub-orbital spaceflights for paying customers. Beyond Branson's Virgin Galactic enterprise, there are other companies currently seeking to take a slice of the potential "space tourism" pie, including Blue Origin, founded by Amazon chief Jeff Bezos, and Bigelow Aerospace, founded by hotel tycoon Robert Bigelow in order to build hotels in Earth orbit. In addition to ferrying paying passengers into space, private space companies are actively exploring such business enterprises as mining asteroids for their mineral content, sending people on vacations into lunar orbit, and exploiting outer space energy sources by, for example, extracting helium-3 from the Moon or building solar-power satellite systems around the Earth. Though it is likely that many of these ventures will fail, they are supported by wealthy individuals with pockets deep enough to overcome many of the research-and-development obstacles they face.

What made all of these ideas—once relegated to the realm of science fiction—feasible? A future historian may look back at the successful 2004 test flights of Rutan's prototype *SpaceShipOne* as the beginning of a new phase of human spaceflight. Before that, no private manned spaceflight had ever occurred, although numerous private citizens have gone into space as part of either the Russian or American space programs. Supporters of the private space industry believe that only the private sector has the vision, finances, and ability to push space exploration to the next level. In the media, the current private space industry has been compared to the aviation industry of the 1920s or the personal computer industry of the 1980s—something embryonic and full of potential. It is, as Glenn Harlan Reynolds notes in his article for the *Atlantic Monthly*, "looking like what a lot of classic—that is, pre-Apollo—science fiction predicted: a slow, steady growth of commercial endeavors in space, with ordinary citizens reaching the moon in the early twenty-first century via the efforts of a wealthy industrialist."

Reynolds's article, "Not Your Father's Space Program," leads off this section with a description of how space enthusiasts are no longer looking to NASA to fulfill their dreams and are growing ever more excited about the possibilities presented by the private space industry. In the next selection, "Space Travel for Fun and Profit," Katherine Mangu-Ward provides a detailed overview of various private space ventures, notably Virgin Galactic, Blue Origin, Bigelow Aerospace, and Space X, PayPal founder Elon Musk's effort to develop a cheap and reusable launch vehicle to take satellites and other space vehicles into orbit. In "Space Business Booming Along the Southwest 'Rocket Belt,'" Leonard David notes that many of today's leading space entrepreneurs are basing their operations in California, Arizona, New Mexico, and Texas. In "Space Race II," Michael Milstein describes how the Google-sponsored Lunar X Prize, which will award $20 million to any private company that can successfully land a robotic rover on the lunar surface, is worrying many scientists who fear that such a mission might disturb the Apollo landing sites, which have been preserved for decades in the vacuum of space. Finally, in "Zero G, Zero Tax," Dennis Wingo advocates using tax incentives to spur the commercial development of privately funded space businesses.

Not Your Father's Space Program[*]

By Glenn Harlan Reynolds
Atlantic Monthly, June 5, 2008

NASA isn't going anywhere very fast these days, but people in the space enthu-siast community aren't as gloomy as you might expect. In fact, they seem more cheerful than I've seen them for years. The reason is that talk about "the space program" (singular) is out of date. Now there are numerous space programs, and while NASA's isn't doing especially well, others seem to be flourishing.

The so-called "space movement" emerged only after NASA's heyday had passed. During the 1960s, when impressive new accomplishments came one after another, those who favored human settlement of outer space were happy just to cheer NASA on. It was only when NASA began to lose momentum following the Apollo program that worried space supporters coalesced into a movement under the umbrella of two organizations: the establishmentarian National Space Insti-tute, founded by Wernher von Braun and Hugh Downs, and the space-hippieish L5 Society, founded by fans of Princeton physicist Gerard K. O'Neill's plan for orbital space colonies. The two groups merged in the 1980s to form the National Space Society, and they've been putting on an annual conference, the International Space Development Conference, for 27 years.

I attended a lot of those conferences back when they were geekfests with an ambience that was half Star Trek convention, half revival meeting. They were fun, and the discussions were often informative, but participants were for the most part on the outside looking in. There was a lot of political discussion, aimed mostly at getting NASA to do the right thing, and at getting Congress to fund NASA to do the right thing, but not a lot of other action.

When my daughter was born in 1995, I quit going for over a decade. I wasn't uninterested, just busy. Recently, though, I attended two conventions—in 2007 and 2008—and discovered that things are different today. From geekfests (not that there's anything wrong with that), the conferences have evolved into something more like the meeting of a professional or trade association. The crowd is better dressed, and often working in the field: one woman I remembered from years back as an activist with Students for the Exploration and Development of Space is now a professor of astronautics at MIT; another former student space activist now coordinates space matters for Google. And truly rich private sector enthusiasts are as conspicuous by their presence now as they were by their absence a decade or two ago.

The main source of excitement at this year's conference was space tourism— folks in the industry prefer the term "personal spaceflight"—involving many companies, from the well-known (Richard Branson's Virgin Galactic) to the sort-of-known (Jeff Bezos' Blue Origin) to the comparatively obscure (XCOR Aerospace). Just this week it was announced that Google co-founder Sergey Brin plans to be a space tourist on a flight to the international space station in 2011. The space tourism business has, according to the results of a study presented at the conference, attracted over 1.2 billion dollars in investment, and grown its revenues from $175 million in 2006 to $268 million in 2007. That's peanuts compared to what NASA is spending on the Space Station or its Constellation project aimed at returning to the Moon, but it's real money and, unlike NASA's budget, it's growing.

Hotel magnate Robert Bigelow is even looking at building his own space station. A (revenue-generating) prototype recently completed its 10,000th orbit around the Earth, and Bigelow has put up $500 million toward making his dream a reality. He'll likely find customers. Indeed, the ISDC featured two wealthy individuals, Greg Olson and Anousheh Ansari, who each ponied up 20 million dollars for a Russian flight to the International Space Station, and both expressed their willingness to pay $100 million for a trip to the Moon if such were available. (Though Olson quipped, "I'd have to sell another company first.")

The Moon is the other main source of excitement. There are now 14 different teams competing for the Google Lunar X-Prize, which will pay $30 million to the first team to send a robot lander/rover to the Moon. Most of them see the prize as a way to jumpstart a longer-term lunar business, not simply as a reward for a one-off effort. Like Virgin Galactic, which followed up on Burt Rutan's win in the original X-Prize competition by creating a space tourism business, it's likely that someone will capitalize on successful private lunar efforts in a more long-term fashion as well.

There's little doubt that most of these ventures will crash and burn, metaphorically if not literally (since this really is rocket science). But because the new ferment in commercial space activity means that NASA isn't the only game in town, it doesn't matter if dozens of space companies fail—so long as some succeed. In this regard, it's an environment more like aviation or automobiles in the 1920s, or computers in the 1970s and 1980s, than like space in the "Right Stuff" era of the 1960s. Back then there was only one player, and it had to succeed or all the effort would have been wasted. In fact, beyond the big names there are now lots of secondary suppliers, who provide components and software to a variety of commercial competitors—a real industrial infrastructure, in other words. It's not a do-the-whole-thing-yourself industry any more.

This trend may also explain the reduced interest in politics that I observed. In previous decades, space supporters were feverishly focused on NASA budgets and electoral maneuverings. Even David Osborne's 1985 cover story for *The Atlantic* on private enterprise in space focused largely on what NASA would do and what Congress would fund. This year, those concerns took a back seat. The convention did feature a well-attended panel on elections, and I hung out in the bar with an impromptu meetup of Obama supporters (their enthusiasm for Obama was unrelated to his views on space, which according to a campaign representative are currently "under review"), but these days no one is looking to the next President to play John F. Kennedy. They'll be happy if the Federal government just leaves space alone.

That's a sign of maturity, too. JFK's decision to go to the moon was a tremendous case of wish-fulfillment for space enthusiasts—instead of a slow growth into outer space, we got a massive commitment of federal resources to get us to the Moon faster than anyone had thought possible. Given that kind of boost, it's not surprising that many spent decades trying to re-enact that kind of success. The JFK moment, however, hasn't come around again, and isn't likely to repeat itself any time soon. And as is often the case when a fantasy becomes reality, the aftermath turned out to be a bit of a letdown; the massive commitment of resources brought about by the political excitement dwindled as soon as that political excitement inevitably disappeared. Politics and command-and-control systems got us to the moon in a decade, but they proved powerless to keep us there. Achieving a space-faring civilization incrementally though profit-making enterprise may take longer, but such an approach doesn't depend on the whims of Presidents, or on right guesses by NASA. It's an approach that takes patience—a quality that has arrived, as it often does, through a combination of time and disappointment.

Ironically, today's state of affairs is looking like what a lot of classic—that is, pre-Apollo—science fiction predicted: a slow, steady growth of commercial endeavors in space, with ordinary citizens reaching the moon in the early twenty-first century via the efforts of a wealthy industrialist. Space supporters, at any rate, will be happy to see this happen . . . especially if they have a shot at buying tickets themselves. And they just might.

Space Travel for Fun and Profit[*]

By Katherine Mangu-Ward
Reason, January 2007

Barbed wire surrounded the Bigelow Aerospace compound, set in a stretch of dry, rock-strewn Nevada desert. Las Vegas glittered in the distance, but otherwise the vista had the desolate look of a lunar landscape, with one difference: The summer heat was oppressive—enough to make you long for the cool vacuum of outer space.

The van full of visiting space geeks didn't mind the harsh conditions. Last July they happily left the air-conditioned glamour of Vegas' Flamingo Hotel and Casino, where the cream of the private space industry had gathered for the New Space 2006 conference, to spend a few hours at Bigelow's warehouse and mission control center. They couldn't have been more excited if the van had been on its way to a *Star Trek*–themed strip club.

Earlier in the week, Bigelow Aerospace had successfully launched Genesis I into orbit. A small pod that inflates once aloft, Genesis I is a prototype for cheap, livable, interconnecting rooms for commercial use in space. The first in a series of launches scheduled every six months for the next two and a half years, it marked the beginning of what could be the first privately funded space station.

Robert Bigelow, president and CEO of the company, made his fortune with the hotel chain Budget Suites of America and other real estate ventures. He has a logical goal in mind: an orbital hotel. Similar in concept to the International Space Station but much larger, Bigelow's space-habitat project uses a cast-off National Aeronautics and Space Administration (NASA) system of inflatable pods. He bought the rights to the technology in 2001, when he read that NASA was scrapping the promising system after many years and many more millions of dollars of development. Bigelow, 62, has since sunk $75 million into the project, with a promise of $425 million more to come.

Stepping inside Bigelow Aerospace's cool, antiseptic, heavily guarded warehouse was like walking into a science fiction novel. Enormous models and pieces of space-bound machinery were strewn about like forgotten Lego blocks over tens of thousands of square feet. The delegation from the NewSpace conference shuffled along with the quiet awe usually reserved for holy places. At one point, a member of Bigelow's mission control team looked at his watch and said, "Actually, Genesis should be passing overhead right now." Everyone in the room looked up, instinctively, as though the module would be visible. Then they grinned sheepishly at each other.

The grins reflected something more than embarrassment at having fallen for an old gag. ("Hey look," someone cracked, "*gullible* is written on the ceiling.") The visitors were just plain happy. After years of hope and speculation, the private-sector space enthusiasts were thrilled to hear the words "It's overhead right now" from one of their own.

The Genesis launch, while exciting, is peanuts compared to what's coming in the next two years. Besides the ever-larger Bigelow launches, scads of private sub-orbital space vehicles will be popping up all over the planet and breaking out of Earth's atmosphere, about 62 miles above sea level.

Bigelow and his ilk are part of an industry that calls itself NewSpace, though some prefer the techy alt.space and others favor the touchy-feely *personal space*. Since the late '90s, they've been coalescing into clubs, nonprofits, and other associations. In the bad old days, this crowd got together mostly to bitch about NASA and its evil stepchildren, Lockheed and Boeing. But while NASA remains a topic of interest, NewSpacers have passed out of their whiny adolescent phase and into industrious young adulthood. Their aspirations are appropriately modest—mostly suborbital, just a quick trip to the edge of the atmosphere. They're setting aside deep space exploration and the moon for now (though they talk a big game about what's next), opting instead for reasonable, practical, short-term goals: quick hops for tourists and other near-to-Earth fun. And instead of crying on each other's shoulders, suddenly the NewSpacers are seeing each other—and sometimes NASA—as the competition.

Thanks in part to a preponderance of tech millionaires, the NewSpace industry is picking up speed. As Bigelow has noted, "We are probably a very close cousin to the world of the Internet and the computer world—doubling every 18 months."

In addition to big-name companies like Virgin Galactic, dozens of smaller entrepreneurial ventures wait in the wings, including Armadillo Aerospace, the rocket company started by *Doom* and *Quake* programmer John Carmack. So do communications equipment manufacturers, spacesuit designers, and many other enterprises, releasing pent-up innovation and creativity as NASA's long-lived monopoly on space, or at least suborbital space, wheezes to an end.

The industry, dominated just a few years ago by a bunch of seemingly loony space cadets with big dreams, is becoming the province of respectable, hardheaded CEOs. What happened?

THREE-HOUR TOURS

The biggest name in the NewSpace business is the British billionaire Richard Branson. The pop entrepreneur founded the space tourism company Virgin Galactic in 2004, and he plans to be flying missions by 2008. Apparently taking a page from *Gilligan's Island*, Virgin will carry paying passengers on three-hour tours, complete with seven minutes of zero gravity, after just a week of preflight training. The Virgin spacecraft will be modeled on SpaceShipOne, the vehicle dreamed up by the aviation legend Burt Rutan. Rutan's spacecraft captured the privately funded Ansari X Prize in 2004 by being the first private manned ship to exit the atmosphere twice in a span of two weeks. After taking the $10 million prize, Rutan's company, Scaled Composites, signed with Branson to build the bigger, better SpaceShipTwo. Rutan says the new ship will fly higher than the first model and carry eight people.

Branson has generated headlines for the private spaceflight industry (and himself) by accepting several $200,000 down payments for early flights. Potential tourist-astronauts include Moby, Sigourney Weaver, Brad Pitt, Stephen Hawking, and Paris Hilton. In March 2005, Doug Ramsberg of Northglenn, Colorado, won a free trip on a Virgin vehicle in a company-sponsored lottery. (Perhaps he'll be one of the lucky few to witness Hawking and Hilton colliding in a brainy yet glamorous zero-g mishap.) Branson says he intends to be on the first flight of the geekily named *VSS Enterprise*, along with members of his family, two years from now.

Two additional companies, Space Adventures and Rocketplane, are also taking reservations and down payments for flights expected to launch on a similar timetable. But they aren't Virgin's only competition in the suborbital sweepstakes.

Elon Musk, the founder of PayPal, now runs SpaceX, which is developing the Falcon rocket series, designed to be a cheap, reusable means of getting satellites and eventually heavier space vehicles for human use into orbit. Falcon testing has been mostly unsuccessful to date, with several delayed launches and an unfortunate fire at the first launch, which sent the rocket crashing into the ocean. But Musk continues to be regarded as a leading figure in the commercial space world—some say the leading figure. NASA recently awarded SpaceX, in partnership with Rocketplane, a $500 million prize to build a vehicle that will deliver crew and cargo to the International Space Station by 2010. Another potential customer for SpaceX is Bigelow, who has expressed a preference for privately developed U.S. rockets. Until a domestic option emerges, all of his modules "are flying on Russian Dneprs," Bigelow says. "They altered and removed nuclear warheads and they're using them for commercial purposes, which I think is pretty damn neat."

The most secretive entrant in the commercial space race is Jeff Bezos, the founder of Amazon.com (and a donor to the Reason Foundation, the nonprofit that publishes this magazine). His space tourism company, Blue Origin, is based on a 165,000-acre Texas ranch. It's developing a rocket ship called the New Shepard. A flight on the vehicle would last just a few minutes and allow a brief period of weightlessness. In mandatory government filings, Blue Origin revealed plans to start testing unmanned vehicles at the end of 2006. Otherwise Bezos has remained mum.

Branson and his peers are confining themselves to suborbital travel for now: blastoff, a few minutes of zero gravity at the edge of space, then back again. The technology to make this type of trip has been around for decades, though NewSpacers are working to make the trip exponentially cheaper, better, and faster. Bigelow's hotel-in-space project is more ambitious, on par with the International Space Station, but also has a longer time horizon. And no one has taken serious practical steps toward a private voyage to the moon, though there has been a lot of discussion about the legal preconditions to make a moon trip attractive to entrepreneurs. For starters, it's not clear how property rights will work on the moon or on asteroids. Who is allowed to build, and where? Perhaps more important, what can be brought back to Earth and sold?

Devotees of private space travel have long blamed NASA's monopolistic behavior for their own failures. And it's true NASA has done virtually nothing to encourage outside innovation over the years—despite repeated mandates to do so—while selfishly sucking up billions of dollars and all the dreams and hopes of space buffs nationwide. But when the NewSpacers lowered their sights from "infinity and beyond" to a few minutes of floating, they realized NASA couldn't really stop them from snagging a little bit of space all their own.

EXTRATERRESTRIAL ENTREPRENEURS

It was 1999 when the free market faction of the space world finally gave up on NASA. In that single year, NASA boasted two failed Mars robot missions, a mostly grounded shuttle fleet, a busted space telescope, and a semi-abandoned space station; it also aborted several pet projects, from a space plane to a planned landing on a comet's nucleus, in large part because they were politically inexpedient. Most space geeks had long ago lost hope that NASA would ever make it back to the moon, as the space agency seemed resigned to sending shuttles scooting back and forth to the International Space Station with small scientific payloads, spare parts, and the occasional astronaut. Pessimists pointed to the average age of NASA professionals, a ripe old 46, and sighed about the lack of innovation. Gone were the Apollo days, when the command was "Waste anything but time." NASA seemed happy to clunk along with its $16 billion a year, doing what it had been doing since the 1970s: not much.

From that despair, the seeds of dozens of companies were tossed to the winds. A few promise bumper crops soon. Once the really big projects were out of the picture—Mars colonies, dinner at the Restaurant at the End of the Universe, etc.—a few guys with big money started to ask: What could be worse than NASA? We might as well try.

Surprisingly, many private space enthusiasts concede that things have started looking up at NASA in the last few years. The Mars rovers *Spirit* and *Opportunity* performed well, pluckily winning the hearts of the American public. The recent shuttle excursion to resume work on the International Space Station got good publicity and actually achieved something of practical worth by delivering new solar panels to the station. And there's another reason the grumbling has diminished: Getting out of NASA's orbit and into their own suborbital groove has left many NewSpacers with a more generous attitude toward the agency, or at least a willingness to turn a blind eye to NASA's failings when their own projects are under way.

The Vegas conference was dubbed NewSpace 2006 but could just as easily have been called "Selling Space," since pretty much everyone in the room was doing just that, in one capacity or another. As one participant noted: "A few years ago, all these guys had the names of struggling nonprofits on their nametags. Today everyone's a CEO."

For years the Space Frontier Foundation, which organized the conference, has been nagging space geeks to stop thinking like engineers and start thinking like businessmen. The trouble with engineers, apparently, is that they are naturally authoritarian. If we could just calculate everything out to the nth decimal place, they say, we could tell you the One Right Way to get to the moon or to launch a rocket. During the no-go '90s, conferences about commercial space ventures were dominated by talk of propellant, rocket design, and lunar habitation specs. "Here's the thing," warns Kevin Greene, founder of a fledgling startup called Lunar Constructors. "There is no 'optimum design' for a moon colony. This is hard for engineers to understand. This is not a libertarian tirade; there is a role for government. But don't over-design it."

Bigelow agrees with the sentiment, adding: "Whether you're building a regional shopping center mall or a 70-story office building, go out and find your anchor tenants. Don't build the whole thing from your idea of what might work."

Having shed their pocket protectors and donned pinstriped suits and silk ties— most of which, mercifully, didn't have little shooting stars or pictures of the starship *Enterprise* on them—NewSpace enthusiasts have grown comfortable with the language, and the indeterminacies, of business. The conference participants talked about "selling ourselves to the public," market segmentation, and strategies to fend off government regulation. Many were starting to think beyond the One Right Way to get to space and beginning to consider extra frills to offer travelers once they're up there.

This isn't to say the field has shed its nerdy image or impulses. Nearly every laptop in the Flamingo's Red Rock meeting room sported space-themed desktop wallpaper. (Perhaps 40 percent were *Star Trek*–related.) One speaker flattered a colleague by saying his "light saber is sharper" than most. And the trade fair presented in conjunction with the conference looked less like a cutting-edge technology expo and more like a middle school science fair. Foam boards with laserprinted pictures of spaceships dotted the walls, and a model rocket engine filled with water bubbled in the corner.

SELLING SPACE

With the recent rise of the market-savvy spaceman, the business plan competition should have been a highlight of NewSpace 2006. Sadly, the viability of the proposals correlated inversely with how awesome they were. Not that rockets aren't cool, but it's hard for the layman to get really excited about which kind of hybrid propellant is going to maximize efficiency.

First among the compelling losers: a pitch for a suborbital football league cum reality show called *Space Champions: Zero Gravity*. The downs, company president Rocky Persaud conceded, would have to be short to fit into the 30 to 40 seconds of zero gravity at the apex of parabolic flight patterns, and tackling in the absence of gravity was going to be tricky. But he optimistically predicted the venture would be profitable almost immediately, with just $3.2 million in financing at the outset. I longed to give this man my money, or someone else's, but his weak PowerPoint presentation and lack of an impressive board of directors suggested he had a long way to go.

Another plan mentioned "memorial space flights" where clients can outdo LSD guru Timothy Leary and *Star Trek* creator Gene Roddenberry by ensuring that all of their cremated remains, rather than just a few grams, make it into space. More measured options included cheap "dedicated launches for small payloads" and easier microsatellite launches. The fun stuff captured the imagination, but the nitty-gritty on rocket manufacture and launch efficiency was bound to win out.

The guys pitching wacky projects have one thing right, though: If the public is going to be interested, it needs to see exciting images and hear wild stories about space. Grainy footage of "One small step . . . " can sustain people's interest only for so long. NASA has lost its touch at selling space, and NewSpace companies are just starting to learn the skill. Virgin Galactic has done the best job so far, with a sharp little product placement in the recent Superman movie: A Virgin Galactic–branded spaceship, possibly piloted by Branson himself, appeared in trailers for the film.

Even without a totally refined message or perfect, snazzy graphics, a handful of wealthy people are ready to get suborbital. The recent, highly publicized trip to the space station by Anousheh Ansari, the entrepreneur who helped bankroll the X Prize, has kindled broad interest in personal space travel. Another female space-traveler-to-be, Reda Anderson, told NewSpace participants she didn't need more reassurance or sales pitches; she preferred the rugged appeal of the young industry. "We're not tourists here," she said. "We're not going to go up and spend time in a hotel and have a nice meal and all that kind of stuff." The first breed of space tourists and entrepreneurs will be attracted, as one conference participant noted, by the fact that space is "fresh real estate, like the Internet," room to grow and expand in an essentially lawless atmosphere (or, more precisely, no atmosphere at all).

But an industry cannot live off adventurers and libertarian dreams alone. Although the market is largely untested, a 2002 survey by the research group Futron found that interest levels were high enough to generate more than 15,000 suborbital tourists by 2021, assuming the price of a ticket comes down to about $25,000 (in 2006 dollars). The Federal Aviation Administration's Office of Commercial Space Transportation put out a report last February estimating the space travel industry would be worth $1 billion a year within 20 years.

The industry is already talking about what's next if and when suborbital jaunts become commonplace. Unlike NASA, commercial space companies answer directly to customer demand, so the dream of pushing on to the moon is strong. "That's what people want—the moon," says Bigelow with a grin. "But we've got a lot of steps before we get there. It doesn't mean we're not always thinking about it, though."

WORKING FOR THE MAN

There's the perennial problem of money, of course—of getting it to the right place at the right time, to ensure that the whole industry doesn't turn out to be a collection of fizzled vanity projects. Bigelow says vaguely that he's had some "serious interest" from customers for short stays in his space hotel, including various governments keen on sending up astronauts and scientists without the expense and hassle of running their own space programs. But Virgin Galactic has $10 million worth of "interest" in hand—in the form of down payments from future space tourists. Virgin spokesman Will Whitehorn has bragged about the company's popularity and its cash on hand, saying, "I'm sure we will have sold out at least the first couple of years by the time we start flying."

The company's first commercial ship is expected to go up in 2008, and Branson says he intends to fly 500 people a year—the same number of people who have been to space in all of history. Perhaps with these confidence-inspiring figures in mind, Virgin inked a deal earlier this year to build a spaceport in New Mexico with some state funding on land set aside by the state for that purpose.

Virgin also has cleared a bureaucratic hurdle that many NewSpacers insist could still kill the newborn industry: the International Traffic in Arms Regulation (ITAR). The problem, oversimplified, is this: U.S. law limits the extent to which you can share technological information in any way related to defense with foreigners. This is understandable when it comes to new kinds of bombs or even snazzy night vision goggles. But it also could restrict the exchange of ideas about propulsion or guidance technologies for purely civilian use. Rick Tumlinson, cofounder of the Space Frontier Foundation and a sort of master of ceremonies at the Las Vegas conference, declared in one presentation: "As far as I am concerned, space ITAR is America's new Iron Curtain. Mr. Bush, tear down that wall."

The partnership between Virgin Galactic and the company that built SpaceShipOne managed to procure an exception to ITAR, though in a way that does not set a strong precedent. The victory came after five months of quiet legal spadework aimed at getting permission for the U.K.-based Virgin to use American SpaceShipOne technology for its commercial suborbital flights. As one attorney discussing the ITAR problem at a conference panel commented, "You should think of lawyers as your friends—your very expensive friends. We want to be part of the solution, not part of the problem."

Meanwhile, the government continues to baffle the young industry by other means. In January 2004, when the White House released its plan to get Americans back to the moon and then on to Mars, even the most hard-bitten NASA cynics took heart in the Bush administration's repeated suggestion that the space agency should contract with private companies for suborbital and orbital services. A mandate for the government to use private contractors where possible would be a serious shot in the arm for the industry. It would also reduce its independence, of course, and would elevate the preferences of government bureaucrats over the preferences of private consumers. Still, the possibility of extra cash was exciting, especially after NASA created the Commercial Orbital Transportation Services program (COTS). It offered a $500 million prize to the company able to come up with a vehicle to move cargo and people back and forth from the International Space Station.

Thirty months later, the Space Frontier Foundation issued a despairing white paper. NASA couldn't let go entirely: While suggesting that it was open to private solutions, NASA invested millions in a redundant system, scaring off would-be contestants for the COTS prize. NASA Administrator Michael Griffin, the paper declared, "claims that he hopes that COTS will work, but that he needs to fund NASA vehicles to go to [the space station] as an 'insurance policy' against the possibility that COTS doesn't work. But most insurance policies are a small amount of the value of the thing being insured."

Part of NASA's unwillingness to gracefully cede suborbital jaunts to the private sector is its fear about "closing the gap." This "gap" is the period between 2010, when the space shuttle is decommissioned, and whenever its replacement kicks in. The COTS prize, recently awarded to a SpaceX/Rocketplane partnership, is encouraging. But it's hard to ask outside entrepreneurs to throw themselves into a project they know will be pushing against NASA's entrenched bureaucracy.

NewSpacers don't consider NASA irredeemable, however. Ken Davidian, a NASA consultant, was the surprise winner of the popularity contest at the Vegas conference. He came representing the agency's prize program, officially known as the Centennial Challenges. Davidian spoke gleefully about competitors for the prizes who spend many more millions to win than the prizes themselves are worth. "Whatever their motivation is, we don't question it very much because we're making out like bandits," he exclaimed. "They do understand capitalism!" whispered Jeff Krukin, executive director of the Space Frontier Foundation, who was sitting next to me in the audience.

But the nerd orgy really began when Davidian clicked over to open a blank PowerPoint slide and asked for ideas for new prizes. For the first time all day, the crowd really perked up. Just a few of the ideas for prize contenders: largest lunar radiation-proof windows, first transmission of one kilowatt of power from orbit, paving the lunar surface for easier landing, and a fully funded college education for the first baby conceived in space. The whole affair was an embarrassment of riches. "This happens everywhere I go," Davidian says. He speculates that "even the most hard-core NewSpace guys really have a soft spot for NASA. NASA got them into this stuff in the first place, right?"

As the conference participants tried to shout each other down, there was a distinct scent of testosterone in the air. Richard Godwin, president of the space publisher Apogee Books, says that's typical of such gatherings. Though a young industry, NewSpace is already subject to some Golden Age-ism. "There was time when it was more purely cooperative, but now there's less collaboration and more competition," Godwin says wistfully. But then he shifts gears: "That's probably a good thing, actually." He even suggests taking the impulse all the way: "Everyone thinks they're the smartest guy in the room. I think we should just have a mud-wrestling match. Everyone, shirts off, and we'll decide who's smartest right here, right now."

The question will be answered one way or another in the next 24 months. Someone will be able to make money by taking people into space on a privately developed, privately owned spaceship. They won't go very far, and they won't be gone very long. But just a few short years ago, the smartest guys in the room were content to sit around and argue better than anyone else. Now—with help from an infusion of smart, *rich* guys—they're fighting for success in a competitive industry with real results on the horizon.

And the mud-wrestling match? You know the industry is growing up, because when Godwin repeated his proposal to several fledgling CEOs, not one reached to unbutton his shirt.

CORRECTION: In "Space Travel for Fun and Profit" (January), NASA's COTS contract was incorrectly described as having been given to a partnership of Rocketplane and Space X. In fact, the award was divided between the two companies.

Space Business Booming Along the Southwest "Rocket Belt"*

By Leonard David
Ad Astra, Summer 2007

The United States has been "belted" for decades. First, there was the Bible Belt—an expression coined in the 1920s by American social scribe H.L. Mencken to describe the American Southeast stretching from Texas to Florida and up as far north as Virginia. And then there was the Corn Belt (Iowa, Indiana, Illinois and Ohio) and the Iron Belt (specifically Northeastern, blue-collar mining states like Pennsylvania), as well as the Rust Belt (Michigan, upstate New York, northern New Jersey, Pennsylvania) and the Sun Belt (the longest belt, stretching from Southern California to Florida).

So why not the Rocket Belt? That term makes sense given the rapid growth of private rocket enterprises that have sprung up to propel public space travel into reality. While Mojave, California, bagged bragging rights as the home of Burt Rutan's Scaled Composites, the innovative aerospace firm that built and launched the first privately funded suborbital spacecraft, SpaceShipOne, farther east, in Texas and New Mexico, small private space companies have also entered the fray.

Historically, each state has been a crucible for space travel. In large measure, Houston was homeport for NASA's budding human spaceflight adventure. In New Mexico, the father of modern rocketry, Robert Goddard, built his prototypes in Roswell in the 1930s. A decade later New Mexico became the official birthplace of space exploration in the United States as V-2 rockets rose skyward from the White Sands Missile Range.

And while private space rocketry hasn't taken root in Arizona to the same degree as in New Mexico or Texas, the state has a preponderance of astronomical observatories, as well as several notable cutting-edge enterprises, such as Paragon Space Development Corporation and its space life support expertise. Then there's the groundbreaking work on the Mars Phoenix Lander at the University of Arizona, in Tucson, as well as observational campaigns of the Mars Odyssey orbiter run from Arizona State University in Tempe. All this demonstrates that space exploration is firmly planted in this region too.

As the 21st century begins with the unfolding saga of public space travel, Texas and New Mexico are arguably in a front-running position to help make a business out of boosting paying passengers to the outskirts of Earth's atmosphere.

CROWD-PLEASER

A notable and enterprising rocket operation in Texas is run by John Carmack, the brainpower behind the video action games Doom and Quake. He has used his good fortune to finance Armadillo Aerospace, in Mesquite, Texas.

Armadillo Aerospace is a leading do-it-yourself business with a research and development team working on computer-controlled rocket vehicles. During the October 2006 Wirefly X Prize Cup, the firm's rocketeers showcased Pixel, their vertical takeoff and landing vehicle. Pixel was a real crowd-pleaser, nearly snagging major NASA prize money in the Northrop Grumman Lunar Lander Challenge.

However, in terms of Texas government interest in his rocket work, Carmack said he hasn't sensed much.

"We have exactly zero interaction with state and local governments, aside from normal business or fabrication permitting," Carmack told Ad Astra. "We might be able to push some slick story about economic development and get some state money, but I don't really go in for that sort of thing. We are here if someone wants to talk about it or do something, but I wouldn't initiate a conversation about it."

Another Texas presence is Space Exploration Technologies Corporation (SpaceX), a company founded by Elon Musk, cofounder of PayPal, the electronic-payment system. He created SpaceX "to help make humanity a spacefaring civilization."

While the privately backed Space X Falcon-class vehicles are being developed and assembled in El Segundo, California, rocket engine and large-scale structural testing occurs in McGregor, Texas. Those booster test stands are bound to be busy this year. SpaceX was one of two winners of the recent NASA Commercial Orbital Transportation Services (COTS) competition. The SpaceX portion of the award was $278 million for three flight demonstrations of the group's Falcon 9 booster, to be topped by their Dragon spaceship. Those tests are scheduled to occur in late 2008 and in 2009. The final flight will culminate in the transfer of cargo to the International Space Station and the return of cargo safely to Earth.

STEPPING-STONE TAKEOFF

By far the most closed-lipped about their rocket plans is Blue Origin. Billionaire Jeff Bezos, of Amazon.com fame and fortune, is digging into his deep pockets to foot the bill on chasing vertical takeoff and landing rocketry. The passenger-carrying suborbital craft they are pursuing is called the New Shepard.

Blue Origin's launch site sits approximately 25 miles north of Van Horn, Texas. It lies within a larger, privately owned property known as the Corn Ranch. Access to the launch site is from Texas Highway 54, which is approximately five miles west of the project's center of operations.

The Bezos planning squad has put in place a vehicle-processing facility, a launch complex and vehicle landing and recovery area. When fully operational, the site will include an astronaut training facility.

On November 13, 2006, Blue Origin rocketeers made their first development flight, a stepping-stone takeoff toward New Shepard. Obviously a proud father in terms of watching the birth of his rocket project, Bezos reported: "We launched and landed Goddard—a first development vehicle in the New Shepard program. The launch was both useful and fun. Many friends and family came to watch the launch and support the team."

Sitting on its landing legs, the large nose-cone-shaped Goddard rocket roared off its launch pad and nosed its way up to roughly 285 feet. Hovering for a brief moment, the vehicle then lowered itself back down to its departure point.

Blue Origin's spaceship work follows, in many ways, the Delta Clipper Experimental (DC-X) and Delta Clipper Experimental Advanced (DC-XA). This Department of Defense-NASA endeavor focused on a single-stage vertical-takeoff, vertical-landing rocket. That ship repeatedly flew between 1993 and 1996 from the U.S. Army's White Sands Missile Range, in New Mexico. Among a list of that vehicle's distinctions—and likely to be part of Blue Origin's milestones—was a 26-hour turnaround that was achieved between the DG-XA's second and third flights—a first for any rocket. The flight program ended in July 1996.

For Bezos and his Blue Origin team, progress over the years is to be measured in step-by-step milestones "to lower the cost of spaceflight so that many people can afford to go . . . and so that we humans can better continue exploring the solar system," Bezos explained.

SECOND SPACE AGE

Without doubt, the X Prize Cup events in Las Cruces, New Mexico, for the past two years have become an epicenter for private rocket businesses.

"Clearly, we are leading the pack in the race to space against states like Florida, Oklahoma, California and Texas," points out New Mexico Governor Bill Richardson. "New Mexico can be very proud of the role we are taking in the Second Space Age."

Richardson has been quick to remark that New Mexico continues to take the lead in advancing the next generation of space vehicles and space travelers. Doing so, he stresses, not only will assist in opening the space frontier to all private citizens "but will bring new companies, provide new jobs, increase tourism statewide and help brand New Mexico as the place to be to experience the future."

New Mexico has already enticed the Rocket Racing league to locate their world headquarters in southern New Mexico. The League is an aerospace entertainment organization that is developing the nascent market for low-altitude, rocket-powered aircraft racing. Similarly, UP Aerospace has established a New Mexico presence to offer their suborbital SpaceLoft XL rocket to toss customer payloads into space—be it for business or educational spaceflight opportunities—and to fit any budget.

In December 2005 Richardson and Sir Richard Branson, chairman of the Virgin Companies, announced that Virgin Galactic would station its world headquarters and Mission Control in New Mexico. The agreement between the state of New Mexico and Virgin Galactic calls for New Mexico to build a $225 million spaceport—now labeled as Spaceport America—in the southern part of the state.

Agreements inked late last year enable the New Mexico Spaceport Authority to secure long-term access to 18,000 acres for Spaceport America with the State Land Office, Sierra County and two private ranch operations. "These agreements are critical to our ability to build Spaceport America and to build a new industry in New Mexico," observed Lonnie Sumpter, executive director of the New Mexico Spaceport Authority.

At Branson's United Kingdom-based Virgin Galactic, ticket sales for suborbital flights at $200,000 a seat are tallying up. As a commercial spaceline operator, Virgin Galactic's first vehicle design of choice—SpaceShipTwo—is now under development at Scaled Composites in Mojave, California. Meanwhile, in New Mexico, the design of Spaceport America is underway to handle the space-bound public traffic by 2009 or 2010. The business plan is for 50,000 people to visit space over a 10-year time period, says Alex Tai, vice president of operations for Virgin Galactic.

ROCKET CITY

New Mexico's support of private rocket developers has also tugged on the aspirations of the UK's Starchaser Industries Inc. That company intends to open the first phase of its New Mexico-based Starchaser Rocket City resort this year.

As the first private space company to establish itself in New Mexico, Starchaser has received welcome support and encouragement from the state, explains Steven Bennett, leader of the rocket company. He told *Ad Astra* that Rocket City will include the opening of a permanent rocket exhibition area and a space-theme hotel with conference facilities, as well as theme restaurants and shops. An office complex and rocket manufacturing plant that hosts public tours are also to be built as part of this first phase. The company has already begun shipping equipment and materials from the UK to its 120-acre New Mexico development site adjacent to Interstate 10 about 17 miles west of the city of Las Cruces.

Starchaser is currently working toward an initial public offering (IPO), which will be used to finance subsequent expansion of its New Mexico operation to include regular space tourism flights aboard its Thunderstar passenger carrying spaceships. Flights aboard the 17-ton, 85-foot-tall Thunderstar-Starchaser 5 rocket vehicles could take place from Spaceport America as early as 2009.

Starchaser is also forging ahead with plans to launch the first of its family of Skybolt sounding rockets in 2007. This totally reusable, unmanned rocket will be flown initially to relatively low altitudes to validate recovery system, telemetry and guidance operations. Although Spaceport America remains a strong contender for hosting these first launches, sites in other parts of the U.S. as well as in Europe are also under consideration, Bennett notes.

So be it blastoffs out of New Mexico or from neighboring Texas, enterprising space concerns are shaping a true Rocket Belt within the United States. Just how vibrant this activity will become over the next few years is hard to forecast. Perhaps a safe prediction is to recall the words emblazoned on Blue Origin's test rocket, the Goddard: "Gradatim Ferociter." That's Latin for step-by-step, by degrees—and fiercely doing so with spirit.

AMERICA'S ROCKET BELT

1 SCALED COMPOSITES—Mojave, California: Burt Rutan's aerospace renegades have changed the game with innovative airplane designs and, of course, the creation of SpaceShipOne.

2 XCOR AEROSPACE—Mojave, California: Rocket engine developers extraordinaire, and aviation innovators creating the power behind the Rocket Racing League.

3 MOJAVE INLAND SPACEPORT—Mojave, California: High up in the California desert, a place where history is made over and over again.

4 BIGELOW AEROSPACE—Las Vegas, Nevada: Building inflatable space habitats and laboratories.

5 ARIZONA STATE UNIVERSITY—Tempe, Arizona: Mission control home for NASA's Mars Odyssey orbiter, now circling the Red Planet.

6 UNIVERSITY OF ARIZONA—Tucson, Arizona: NASA's next Mars mission, the Phoenix lander, will be operated from the university's campus.

7 PARAGON SPACE DEVELOPMENT CORPORATION—Tucson, Arizona: Specializes in thermal control and life support in extreme environments, deep in designing key hardware for human-rated spacecraft.

8 SPACEPORT AMERICA—Las Cruces, New Mexico: Home of the annual X Prize Cup festivities, as well as the Rocket Racing League, Up Aerospace, Starchaser Industries, and Virgin Galactic space-liner operations. Work is underway on the first "purpose built" spaceport to handle an array of entrepreneurial access to space projects.

9 BLUE ORIGIN—Corn Ranch, Texas: Amazon.com leader Jeff Bezos is bankrolling the New Shepard, a vertical-takeoff, vertical-landing vehicle program designed to take a crew on a suborbital journey into space from the firm's 165,000-acre launch site.

10 SPACE EXPLORATION TECHNOLOGIES CORPORATION (SpaceX)—McGregor, Texas: Roaring rocket engines can be heard at this SpaceX 300-acre testing facility.

11 ARMADILLO AEROSPACE—Mesquite, Texas: Controlled liquid oxygen-ethanol rocket vehicles, with an eye toward piloted suborbital vehicle development.

12 OKLAHOMA SPACEPORT—Burns Flat, Oklahoma: As the future launch site of Rocketplane, Oklahoma was first state to get behind building a spaceport for private spaceflight.

Space Race II[*]

By Michael Milstein
Smithsonian, June 2008

The second race to the moon has begun—and this time there will be a big cash payout for the winner. Four decades after Neil Armstrong took his giant leap for mankind, the Google-sponsored Lunar X Prize is offering $20 million to any private team that puts a robotic rover on the moon, plus $5 million in bonus prizes for completing such tasks as photographing one of the numerous man-made artifacts that remain there—for instance, the Apollo 11 lunar module descent stage that Armstrong and Buzz Aldrin left behind in 1969.

One goal of the Lunar X Prize is to rekindle excitement in space exploration by beaming pictures of historic lunar locations to Web sites or even cellphones. But dispatching robots to snoop around the moon also poses a risk to some of the most precious archaeological sites of all time. What if a rover reached Tranquility Base, where Armstrong landed, and drove over footprints, which are still intact and represent humanity's first expedition to a celestial body? William Pomerantz, the director of space projects for the X Prize Foundation, acknowledges that possibility. "There's always a tradeoff between wanting to protect the history that's already there and wanting to visit the history," he says.

The competition brings into focus a potential problem that worries a growing circle of archaeologists and space historians: the careless destruction of invaluable lunar artifacts. At Charles Sturt University in Australia, Dirk H. R. Spennemann—who specializes in the preservation of technological artifacts—says Tranquility Base symbolizes an achievement greater than the building of the pyramids or the first Atlantic crossing. And because the moon has no atmosphere, wind, water or known microbes to cause erosion or decay, every piece of gear and every footprint remain preserved in the lunar dust. Spennemann advocates keeping all six Apollo sites off-limits until technology enables space-faring archaeologists to hover above them, Jetsons-like. "We only have one shot at protecting this," he insists. "If we screw it up, it's gone for good. We can't undo it."

The initial response to the Lunar X Prize initiative—which had ten registered teams at the end of April—suggests the moon's remoteness won't discourage unofficial visitors for long. History teaches a similar lesson. When the *Titanic* sank in 1912, few imagined that it would become an attraction. But not long after Robert Ballard discovered the wreckage in 13,000 feet of water in the North Atlantic in 1985, treasure hunters in submarines looted the doomed vessel of jewelry and dinnerware.

Crafting an agreement that bars exploration of lunar sites in the coming age of space tourism may be difficult. To be sure, nations retain ownership of spacecraft and artifacts they leave on the moon, though it (and the planets) are common property, according to international treaties. In practical terms, that means no nation has jurisdiction over the lunar soil, upon which artifacts and precious footprints rest. "It would be our strong preference that those items remain undisturbed unless and until NASA establishes a policy for their disposition," says Allan Needell, curator of the Smithsonian National Air and Space Museum's Apollo collection. The "preservation of the historical integrity of the objects and the landing sites" would be a primary goal, he adds.

How much stuff have people left on the moon? Professors and students from New Mexico State University (NMSU) cataloged equipment left behind at Tranquility Base and identified more than 100 items and in situ features from Apollo 11 alone, including Buzz Aldrin's boots, Armstrong's famous footprint and a laser ranging retroreflector, which, for the first time, measured the precise distance between the moon and Earth. Much of the equipment was discarded by Armstrong and Aldrin just prior to lifting off to rendezvous with the orbital craft that would take them home; they needed to lighten the lunar module ascent stage, which they'd burdened with 40 pounds of lunar rocks and soil.

The New Mexico researchers had hoped that their inventory would help them gain protection for Tranquility Base as a National Historic Landmark. But the National Park Service, which oversees the program, rejected the proposal, saying the agency doesn't "have sufficient jurisdiction over the land mass of the Moon." Moreover, a NASA lawyer advised that merely designating a lunar site a landmark "is likely to be perceived by the international community as a claim over the Moon"—a land grab that would place the United States in violation of the 1967 Outer Space Treaty. So Beth Laura O'Leary, an anthropologist who led the NMSU project, added the historic lunar site to an official list of archaeological sites maintained by the state of New Mexico. It's a largely symbolic gesture, but it does mean at least one governmental body recognizes Tranquility Base as a heritage site. "You don't want people putting pieces of Apollo on eBay any more than you want them chiseling away at the Parthenon," O'Leary says.

Of course, NASA itself has done some extraterrestrial salvaging. In 1969, in arguably the first archaeological expedition conducted on another world, Apollo 12 astronauts Alan Bean and Pete Conrad visited the robotic Surveyor 3 spacecraft, which had landed two years earlier. They inspected the landing site and removed the spacecraft's television camera, a piece of tubing and the remote sampling arm.

The parts were returned to Earth so researchers could assess the lunar environment's effects on equipment.

While archaeologists take a hands-off approach to the six Apollo landing sites, researchers are more open to granting access to robotic sites. Charles Vick, a senior analyst at GlobalSecurity.org and an authority on the Russian space program, says historians could learn a lot about the still-shrouded Soviet space program by studying equipment left behind during the USSR's Luna probes, which landed between 1966 and 1976. In 1969, the USSR's Luna 15 probe crashed into the moon. Its mission was believed to be collecting lunar rocks and returning them to Earth, but scholars in the West still aren't sure. "We're not going to know until we go there and check it out," Vick says.

Without new international agreements, the norms governing lunar archaeology are likely to remain vague. The Lunar X Prize rules state that an entrant must get approval for a landing site and "exercise appropriate caution with regard to the possibility of landing on or near sites of historic or scientific interest." Teams going for the bonus prize must submit a "Heritage Mission Plan" for approval by the judges, "to eliminate unnecessary risks to the historically significant Sites of Interest." (Lunar X Prize participants were scheduled to meet in late May to discuss the rules and guidelines.) Still, the contest rules don't specify what constitutes an unnecessary risk. And there's no guarantee where the competing spacecraft will end up. With no traffic cops on the moon, the only deterrent against damaging sites might be the prospect of negative publicity.

O'Leary says the Lunar X Prize's lack of regulation is "scary"—a sentiment shared by others. But at least one Lunar X Prize entrant, William "Red" Whittaker, a professor of robotics at Carnegie Mellon University, has a simple solution to minimize risk: after landing, his team's rover would use telephoto lenses to view Tranquility Base from afar.

To Pomerantz, the competition's director, merely debating how to protect lunar history is a welcome sign that humanity is finally on the verge of going back: "It's exciting when questions that seemed distant and hypothetical are becoming not too distant and not too hypothetical after all." For now, archaeologists are just hoping a robotic rover doesn't take a wrong turn.

Zero G, Zero Tax[*]

Enabling Private Space Enterprise Through Tax Incentives

By Dennis Wingo
Ad Astra, Spring 2007

Money! Money is the true rocket fuel of the commercial Space Age. There is a quote in the movie The Right Stuff that one of the astronauts getting ready for the Mercury mission says: "No bucks, no Buck Rogers!" This statement has become an axiom in both government and commercial space. No bucks killed the Apollo program when the money was shifted away from space in the early 1970s. The space shuttle's development was constrained by money, and the vehicle we have today is the result. The space station? Same story. How about the first Space Exploration Initiative under President George H.W. Bush? No bucks, no lunar base! These are the facts of life regarding space.

With the Congress telling NASA it must live with the same budget that they had last year, the first ominous clouds of future lack of congressional interest in the Vision for Space Exploration are on the horizon. If this is the case, then what can be done by our community to support private space efforts?

Since the dawn of the Space Age, there has been very little movement toward private markets for human spaceflight NASA has always been the 900-pound gorilla in that market, and the perception has always been to avoid the market like the plague by the investment world. It took the risk capital from Paul Allen, who funded the technically brilliant Burt Rutan, to build the first prototype of a mass-market suborbital space tourism system. However, Allen probably would not have invested had there not been some form of return in the form of the Ansari X Prize. Allen is what we in the space business call a visionary investor, or VI. A VI, while looking for an economic return, does have an agenda beyond pure return on investment. Yet the vast majority of the risk capital out there sees human-related space as too great of a risk, with too little return to invite their participation.

We cannot tell today what will be the "killer app" of the commercial Space Age.

However, we do know what works to enable the growth of new industries. There is a long history of this in the development of America as a capitalist society. It was a mixture of a granted state monopoly and private investment that enabled Robert Fulton to build the world's first practical steamship. It was a mixture of government bonds and private investment by names like Stanford, Huntington and Crocker that enabled the nation to be united by the transcontinental railroad.

In 2000 the first serious effort to address this issue was undertaken with the introduction of the Zero G, Zero Tax (ZGZT) legislation. Here is a short history of the legislation provided by Alex Gimarc as part of a white paper on the subject:

- First introduced in 2000. Provided for 20-year tax holiday for new space products and services. To attempt to maintain revenue neutrality, existing profitable industries were excluded; thus the definition of eligible products and services excluded "any telecommunications service, any service provided by a weather or other Earth observation satellite and any service of transporting property to or from outer space."
- Reintroduced in 2001. Exclusions changed to "any telecommunications service provided from Earth orbit, any service provided by a weather or other Earth observation satellite, and any other service provided on or before the date of the enactment of this section of transporting property to or from outer space."
- The 2005 version incorporates some tax-credit concepts from the former Calvert-Ortiz tax bill (i.e., Invest in Space Now Act). Can be seen as merger of two bills.

The ZGZT legislation has had many sponsors and actually almost passed in the House of Representatives in 2001. The bill failed because the congressional budget office examined the tax consequences of the bill at $10 billion over its 20-year life. This was not examined for its positive aspects. Today there is zero revenue by any company that would be covered under the ZGZT legislation. In order to cost the government $10 billion, the companies have to make a profit of $28.57 billion over that time period (assuming the standard 35% corporate tax rate). If we use a conservative 10% profit margin for these companies, this implies that the aggregate revenue over the 20-year period is $285.7 billion!

Let's take this a little further. It is typical for companies in the high-tech engineering world to have a cost of labor of 30–50% of revenue. This means that the salaries for all of the people who work for the ZGZT-enabled companies are between $86 billion and $143 billion. Most of these folks have mid to high-paying jobs, meaning that we can take a conservative 18% of their salaries for federal taxes and 15.3% for social security and Medicare taxes. This brings a total tax revenue into the federal treasury as follows:

What the table [below] clearly shows is that even using very conservative numbers for salaries as a percentage of revenue and taxes as a percentage of salaries, the net gain to the federal treasury is between $18 billion and $37 billion over the life of the bill. This is pretty good for an industry that did not exist before the passage (potential) of the bill. This is called dynamic scoring in congressional legal terms, something that Congress did not do when they considered the ZGZT bill previously. It is this type of argument that has to be made for Congress to really understand how this bill enables space commerce.

What about the investor? The investment community well understands the effect of tax policy on the growth of industry. The tax holiday on the Internet was one of the crucial factors enabling its growth from a few hundred academic computers in the 1980s to the global force that it is today. This is also the potential for space. We as space advocates know the value of opening the solar system for economic development. We have not done a good job over the years in communicating this vision. We have an opportunity with ZGZT and similar legislation to let dollars speak for us with the result that Buck Rogers takes on a whole new meaning!

Table 1: Salaries as a Percentage of Total Revenue and the Resulting Tax Consequences

	Salaries (Billions)	Fed Taxes (18%) (Billions)	SS & Medicare (Billions)	Total Taxes (Billions)
30% Revenue	$86.71	$15.43	$13.31	$28.54
40% Revenue	$114.28	$20.57	$17.48	$38.05
50% Revenue	$142.85	$25.71	$21.85	$47.57

4

Roving the Red Planet:
The Continuing Exploration of Mars

The thin Martian atmosphere as seen from Viking.

Editor's Introduction

Mars, the fourth planet in our solar system, has long captivated humanity. Part of the attraction is likely due to its visibility to the naked eye, the blood-red color it emits in nighttime skies, its tantalizing *nearness*. It held a particular fascination for the people of ancient Rome, who named the planet after their god of war—the son of Jupiter and Juno and one of the most revered gods in their pantheon. The Romans considered themselves descendants of Mars, who was said to be the father of the city's legendary founders, Romulus and Remus. Many ancients also believed that the rising and setting of the planet Mars, which they thought was a star, portended the future. Today, many people still imagine that the red planet foretells our future—only now as a place where the human race will one day settle and live.

Scientific observations of Mars have been ongoing for centuries. The 19th-century Italian astronomer Giovanni Schiaparelli used his telescope to study Mars in great detail and quickly became convinced that he saw a wide network of canals, of either the natural or artificial variety, crisscrossing the surface. Percival Lowell, a Boston millionaire long fascinated with astronomy, was certain these canals were the work of alien intelligence. Lowell spent years in his observatory in Arizona studying these canals, which he believed were evidence that the fourth planet was populated by an ancient civilization that had constructed them to bring water across their dying planet from the polar icecaps. Other astronomers came to believe that the apparent changes in color on the Martian surface suggested the planet had seasons similar to those on Earth, with plant life blooming in warmer months and lying dormant in the winters.

It was not until the advent of space exploration that the concept of a living Mars supported by Schiaparelli, Lowell, and a host of sci-fi writers, was finally put to rest. In July 1965 the U.S. probe *Mariner 4* sent back the first photos of Mars, which revealed a dry and barren world—not a place that could have supported life in the recent past. Since that time dozens of spacecraft, including orbiters, landers, and most recently rovers, have been sent to Mars by the American, Russian, European, and Japanese space agencies. In their studies of the Martian climate, geology, and surface, these probes have revealed findings that strongly suggest the planet was once a far warmer and wetter place—the kind of setting that may have harbored the elements needed for the evolution of life. While most scientists dismiss the idea that an advanced civilization could have ever existed on Mars, many are nevertheless intrigued by the notion that some sort of primitive life could have evolved there.

Though these robotic probes have thoroughly mapped and studied Mars, this exploration has not always proceeded without difficulty. An interesting side-note to Martian exploration is the fact that roughly two-thirds of the spacecraft bound for Mars have failed to complete their missions. Many of these aborted missions resulted from obvious technical glitches, but a number have failed for unknown reasons—leading some to playfully postulate that there is a "Mars curse," or as one NASA in-joke suggests, a "Great Galactic Ghoul" that eats Mars-bound spacecraft. Of all the Mars exploratory programs, the American program has had the greatest success. As of this writing, NASA has successfully launched 13 out of 18 of its missions and placed six out of seven of its landers on the Martian surface—an impressive record when one considers that only half of the 38 total Mars-bound launches ever reached their destination.

Because of NASA's success rate and its aggressive robotic exploration of the planet since the 1990s, the articles in this section focus primarily on the U.S. Mars program, though an overview of all Martian exploration is provided. In the first entry, "Mariner IV to Mars and More," Katherine Bracher presents a short history of the exploration of Mars since the 1960s. In the next piece, "Inside the Mars Rover Missions," Bill Farrand discusses the enormously successful (and ongoing) NASA rover missions. The twin rovers, *Spirit* and *Opportunity*, were supposed to traverse the Martian surface for a hundred days after landing in 2004; they have since far exceeded their design specifications and continue to provide detailed information to scientists. In "What We Learned on Mars," Ivan Semeniuk cites cleansing Martian winds, extensive pre-launch testing, and close collaboration between NASA scientists and engineers as the main reasons behind the success of the rover missions. David Brown, in "Lander Finds Ice on Mars, Scientists Say," reports on the current Phoenix mission to the red planet, which spotted evidence of sublimation at the Martian North Pole. Finally, Carl Zimmer, noting the difficulty of recognizing fossilized microbes here on Earth, questions how we will do so on Mars in his article, "Life on Mars?"

Mariner IV to Mars and More [*]

By Katherine Bracher

Mercury (San Francisco, Calif.), July/August 2006

Many of us have lately seen the IMAX movie about Spirit and Opportunity, the recent highly successful Mars rovers. The film provides spectacular coverage of the development of the mission, the launches, the landings, and some of the results of the rovers' explorations. But some of us still remember our excitement forty years ago, when we saw the first grainy images of the Martian surface from the U.S. space probe Mariner IV. Carl Sagan described this mission in an A.S.P Leaflet for July 1966, and I wrote about it in an "Echoes of the Past" column exactly fifteen years ago. Since then much more exploration of the Red Planet has given us a very different view of Mars than we had in 1966, or even in 1991.

Mariner IV flew to within 14,000 km of Mars, sent back 22 pictures, and collected data on the Martian atmosphere and magnetic field. Sagan wrote: "Although these 22 photographs covered only about 1% of the Martian surface, they have made a most unexpected and amazing contribution to the history of planetary exploration." The biggest surprise was the presence of large numbers of craters, closely resembling those on the Moon. Sagan continued, "[t]hese probably derive from impacts from the asteroid belt . . . The presence of fairly large craters and the absence of obvious signs of water erosion suggest that there must have been no extensive bodies of liquid water on the Martian surface during recent epochs." At that point the conclusion was that Mars was probably much like the Moon, with a heavily cratered surface and little other activity.

In the next few years Mariners 6 and 7 flew by Mars, giving us views of other parts of the planet, and in 1971 Mariner 9 went into Martian orbit. A major dust storm was raging when it arrived, and only after several months did the spacecraft collect images that showed some apparent riverbeds and channels, and a glimpse of the enormous volcano Olympus Mons. In 1976, the Viking 1 and 2 spacecraft made soft landings on the Martian surface. They provided views of terrain littered with rocks, sand dunes, and a pink sky on the horizon. The crafts' orbiters also showed much more surface detail than previous missions. It became clear that, as I wrote in 1991, "large volcanoes like Olympus Mons, huge canyons like Valles Marineris, and complex erosion and dune patterns that suggest past liquid water, indicate much more geological activity" than anyone had expected.

Following the Viking missions, a hiatus of some twenty years ensued before another successful U. S. Mars probe. In 1993 the Mars Observer failed, but in 1996 the Mars Global Surveyor went into orbit. In 1997 the Mars Pathfinder made a soft landing, extended a ramp to the ground, and the Sojourner Rover rolled down onto the surface and began 83 days of roaming about. Sojourner returned over 500 images of the terrain in its vicinity, and captivated viewers who eagerly followed its tracks across the ground.

The nation's next several attempts to explore Mars met with failure—these included the Mars Climate Orbiter, Deep Space 2, and the Mars Polar Lander. More recently, other craft have gone into Martian orbit: Mars Odyssey in 2001, the Mars Exploration craft with their rovers in 2003, and the European Space Agency's Mars Express in 2003. The latest is the Mars Reconnaissance Orbiter, which entered orbit around Mars in March of 2006. It is designed to look for underground deposits of water, among other things. Meanwhile, over two years after the rovers Spirit and Opportunity began their treks on opposite sides of Mars, they are still exploring and sending back data.

Imaging technology has developed amazingly over the past forty years. Where the Mariner IV images were grainy and of very low resolution, today's images are sharp and detailed, and can be manipulated to give stereoscopic and 3-D views of regions of Mars. The IMAX film shows some of this, and you feel as if you are there.

The picture of Mars emerging from all these observations is one of an active planet. The Mars Express has provided evidence that active volcanism occurred quite recently (a mere few million years ago). It also appears that there are glacial deposits, and glaciers may have advanced and retreated with changes in Mars's orbit. It has been suggested that one flat region on the Martian surface is a frozen sea of water ice; others think it is lava. Clearly much remains to be learned about this neighboring world.

Inside the Mars Rover Missions[*]

By Bill Farrand
Ad Astra, Winter 2005

I think the longest 11 minutes I ever experienced in my life were those 11 minutes on the evening of Jan. 3, 2004. I, along with other members of the Mars Exploration Rover science team, were assembled in our science operations working group meeting room at the Jet Propulsion Lab in Pasadena, Calif., awaiting the landing of the rover Spirit in Gusev Crater. We had been listening with anticipation as the mission engineers reported on Spirit's progress in its passage through the Martian atmosphere.

Described to the press as "six minutes of terror," Spirit apparently had survived that perilous journey with flying colors. In fact, the rover had just sent back a signal saying that its airbag landing system had deployed and that it was bouncing, but then . . . silence.

Heavy on all our minds were the recent failure of the Beagle 2 lander and the twin failures in 1999 of the Mars Polar Lander and the Mars Climate Orbiter. We had all invested so much time and energy in these two rovers that to lose them was unthinkable. So we waited, and waited, and then . . . Success! A signal was received that Spirit had rolled to a stop and was safe. An adventure that science principal investigator Steve Squyres of Cornell University in New York has called "the first great mission of exploration of the 21st Century" was ready to begin.

And what an adventure it has been for those of us who have been working with the rovers on a daily basis. While living and working on Earth, we were working and sleeping on the same time scale that the rovers followed. This meant extending our days by 39.24 minutes to match the length of the Martian day. So we would sometimes sleep during the sunlight hours on Earth, come into JPL and greet our co-workers with a chipper "good morning" at 9 p.m. Pacific time.

Checkout of Spirit's systems proceeded apace and we were delighted with the images of the interior plains of Gusev Crater that were spread out around Spirit and its landing platform. Much to our surprise and delight, visible on the horizon were the hills that we would come to call the Columbia Hills in honor of the Columbia space shuttle astronauts. The hills seemed tantalizingly close and yet frustratingly far away. The mission success goal for the rovers was for each of them to drive 600 meters, or 1,968 feet, and the hills were over two kilometers, or 1.24 miles away. Little did we know at the start of the mission how sturdy Spirit would prove to be.

For a heart-stopping series of days, though, we feared that Spirit's mission might end before it really began. Just as Spirit was poised to make its first in situ measurements on the pyramidal rock "Adirondack," it suffered a serious "anomaly"— "anomaly" being engineer-speak for an "oh no!" problem—which proved to be caused by Spirit "overfeeding" on data and loading its flash memory past capacity. The engineers worked through the problem, but as they struggled with Spirit's glitch another small issue had to be dealt with—the landing of Spirit's twin rover, Opportunity.

LUCKY LANDING

On Jan. 24, Spirit's picture-perfect landing was mirrored by the equally successful landing of its sister on the flat plains of Meridiani Planum, located on the other side of the planet. While the first images we had received from Spirit had made us happy, those we received from Opportunity made geologists such as myself ecstatic because in an extraordinary "hole in one," Opportunity's airbag bounce and roll had deposited it in a small 22-meter (72-foot) diameter crater. Even better, there in front of the rover was in-place bedrock. All previous Mars landers had seen plenty of rocks to be sure, but they had all been out-of-place rocks transported from their source regions by impacts or floods. As we studied the images of the "Opportunity Ledge" outcrop that came in from the rover's color Panoramic Camera, or Pancam, and initial thermal infrared measurements of it by its "mini-TES" thermal emission spectrometer, we began to realize that there before Opportunity was what we had come to Mars to find: rocks that by all initial indications were sedimentary in nature.

Sedimentary rocks on Earth come in a variety of forms and textures and often volcanic ash beds can mimic many of the textures and bedding patterns of water lain sedimentary rocks. We had a good deal of discussion about whether the rocks in the Opportunity Ledge outcrop were in fact produced by the action of water or were volcanic ash beds. The only way to find out for sure was to go up to them and have a close-up look just as any Earth-bound geologist would do with any mysterious rocks he might find in the field. Before reaching the outcrop, Meridiani Planum had another surprise in store for us. After rolling off its landing platform and taking a closer look at the floor of the crater with the Pancam, we found that the crater floor (and later we would find the surrounding plains as well) was blanketed with small (on average five millimeters in diameter) spherules. In the color Pancam images they appeared blue and were promptly dubbed as "blueberries."

The mystery of the blueberries took a back seat to the allure of the Opportunity Ledge outcrop. A field geologist will scratch or break a rock to obtain a fresh surface and take a close-up look at the rock texture with a magnifying lens. Opportunity and Spirit were equipped to do much the same thing. Hence when Opportunity made its first measurements of the mysterious light-toned rocks exposed in the inner walls of Eagle Crater, it took a close-up look with the rover's Microscopic Imager, obtained a fresh surface by grinding into the rock with the Rock Abrasion Tool (RAT), and "sniffing" the rock—examining its chemical and iron-bearing mineral composition—with its Alpha Particle X-Ray Spectrometer (APXS) and Mössbauer spectrometer.

The first up-close views of the outcrop indicated that the blueberries were, in fact, weathering out of the outcrop. Moreover, that rock provided yet another set of surprises. First, chemical analysis of the outcrop by the APXS indicated that it was richer in sulfur—up to 40 percent sulfate minerals—than any rock or soil yet analyzed on Mars. The first measurements by the Mössbauer spectrometer provided perhaps an even bigger surprise since they indicated that the outcrop was filled with the mineral jarosite, a hydrated iron sulfate mineral that, on Earth, always forms in the presence of water. So, to use mixed sports metaphors, Opportunity's "hole in one" at Eagle Crater led to a "home run" with its discovery of rocks formed in water.

SEASONED TRAVELER

While Opportunity was busy analyzing the rocks of the Opportunity Ledge outcrop, Spirit, after having been cured of its case of "over eating" on data, was proving to be the long distance voyager of the twin rovers. It rolled across the plains of Gusev Crater to the nearly 200-meter (656-foot) diameter Bonneville Crater, stopping along the way to do its own analysis of rocks and soils. What Spirit found were dark rocks with the composition of basalt—no big surprise there since the presence of basalt on the surface of Mars had been indicated by telescopic observations and measurements by the orbiting Thermal Emission Spectrometer (TES) on-board the Mars Global Surveyor spacecraft. The soil also had an overall basaltic composition, but the Mössbauer spectrometer indicated that it had abundant olivine, a mineral that rapidly breaks down in the presence of water.

So while Opportunity had found evidence of a possible past sea, Spirit had found a disappointingly dry desert. While we had hoped that Bonneville Crater might have outcrops of bedrock, but upon reaching it, imaging by the rover's Pancam and remote spectroscopic measurements by its mini-TES indicated that it appeared to be a crater in a big rubble pile with no in-place rock outcrop. While they seemed frustratingly far away, hope for finding evidence of a past lake in Gusev Crater now seemed to rest entirely on Spirit reaching the Columbia Hills. More than one of my colleagues would say, "We've got to get to those hills!"

SCIENTIFIC PAYDIRT

While Opportunity's examination of the rocks outcropping in Eagle Crater had confirmed the presence of jarosite, the multispectral Pancam imagery also indicated the presence of red hematite, the finer grained cousin of the coarse grained gray hematite that had been the reason that Meridiani Planum was chosen as a rover landing site. The host of the gray hematite, detected from orbit by the TES instrument, remained a mystery until Opportunity was able to take a close-up look at a concentration of "blue berries" in a feature named, appropriately enough, "the Berry Bowl." Measurements by the spectrometers onboard Opportunity indicated that the berries were composed largely of gray hematite. This seemed to indicate that they were actually "iron concretions"—nodules formed by the secondary circulation of fluids within sedimentary rock outcrops. While such concretions are not necessarily rare on Earth, they are not found in the abundance that is apparent on the plains of Meridiani Planum.

Initial examination of the Opportunity Ledge outcrop in color and high-resolution stereo imagery had revealed distinct layering. It took closer examination, and a search for the best examples, to confirm that in addition to minerals formed in water (which quite plausibly could have formed by the circulation of ground waters), the layers in these rocks formed distinct cross beds in places. Cross bedding is a form of sedimentary structure produced by the movement of sediments by water or winds. The cross beds observed in the Opportunity Ledge outcrop had structures deemed distinctive of movement by water—confirming to many of my colleagues that these rocks had formed in a long-gone standing body of water and were, in fact, evaporates. Evaporates are rocks formed by the evaporation of ancient lakes or seas and the resulting concentration of salts and related minerals into sediments and consequently, rocks.

In essence, Opportunity had succeeded in its mission—finding evidence of the sustained action of water at the surface of Mars. Thus, potentially, finding evidence of an environment where life would have had the ingredients, and the time, to come into being and, for a time, prosper.

While the explorations of Opportunity of Eagle Crater had been fruitful, other parts of the Meridiani plains beckoned. Thus after 56 Martian days, or sols, exploring Eagle Crater, Opportunity emerged from the crater, started across the plains to the larger (150 meter in diameter) Endurance Crater which lay 800 meters to the east of Opportunity's landing site.

After 39 sols of roving across the exceedingly flat plains of Meridiani Planum, with several significant science stops along the way, the rover reached the western rim of Endurance Crater (informally named in recognition of Antarctic explorer Ernest Shackleton's ship) and was rewarded with a spectacular vista. While Opportunity had spent nearly 56 sols examining only about a meter's thickness of outcrop, the walls of Endurance Crater had more than 10 meters of outcrop.

Opportunity's exploration of Endurance began with a partial circumnavigation of the outer rim of the crater, with two stops for panoramic imaging of the crater interior. At the same time, engineers were performing computer simulations and physical trials with a test bed rover to see if Opportunity could safely descend into the crater (and later drive out) and begin to sample those outcrop layers directly. Eventually, the decision was reached, and a location found, that was deemed safe for the rover to drive in and its sampling of the layering in Endurance began.

A NEW WORLD

Meanwhile, on the other side of the planet, Spirit's long traverse across the plains of Gusev Crater had been successful. Along the way, Spirit had found tantalizing hints of the action of water: rocks with fractures and veins potentially affected by water, rounded pebbles armoring dunes, layered coatings on basaltic rocks. However, no "smoking gun" to compare with the cross-bedded, sulfate-rich rocks found by Opportunity had yet been found by Spirit. Each rock examined by Spirit was found to be variations on a theme of basalt. However, all that was to change when Spirit finally reached the Columbia Hills. Upon reaching the base of the hills, Spirit found a unique set of rocks with an odd "inside out" form of weathering. Analysis of these rocks revealed that it had a generally basaltic composition, but that it also had a mineral common at Meridiani Planum—hematite. These rocks with unique names such as "Pot of Gold" and "Breadbox" have been deemed as very likely being altered by the action of water.

Spirit's journey had been long and hard and while the "promised land" of the Columbia Hills was at hand, Martian winter was approaching and with that change of season, the effects of shorter days and lower sun angles led to the solar powered rover's power situation getting worse and worse. Adding to the difficulties, Spirit's drive had exceeded its tested "warranty" and its right front wheel showed symptoms of overuse. Recent drives have been done backward, using five of the rover's six wheels in order to preserve what is considered a limited number of rotations before the wheel becomes locked in place.

The fact that Spirit has reached the Columbia Hills is actually fortuitous for the energy situation since, by driving up onto north facing slopes of the hills, Spirit is placed in a more favorable angle for solar charging given Spirit's location in the southern hemisphere. With the improved energy situation Spirit is poised to continue its exploration of the Columbia Hills. That exploration continues to be interesting because as of Sol 190 of the mission, Spirit had finally matched its twin rover by finally finding bedrock! While the exact nature of that bedrock still is in doubt, it is clearly not the same type of dark basalt found out on the plains. The material is softer and lighter in color than the volcanic rocks littering the plains. Further examination of the rocks making up the Columbia Hills should provide more information on the history of water in Gusev Crater.

While Spirit had found nothing but dark rocks until reaching the Columbia Hills, the only in-place rocks found by Opportunity have been variations on a theme of the light-toned rocks first seen in Eagle Crater. However, in situ examination of the layers making up the walls of Endurance Crater have provided important chemical details that should eventually provide a better picture of the history of water at Meridiani Planum. Opportunity has now left Endurance Crater to continue its exploration of the Meridiani plains.

Spirit and Opportunity have far surpassed the expectations that any of us had before they landed nearly one year ago. They have transformed our view of Mars from single views from isolated locations to real, changing landscapes filled with a surprising variety of rocks, soils and skies sometimes peppered with clouds. They have also given us evidence that Mars was once a much different place with salty, probably acidic, seas. Whether life ever formed in those waters is a question that will have to be resolved by future missions, but the Mars Exploration Rovers have given us the spirit to proceed. Now we must seize the opportunity and continue the exploration of our fascinating neighbor in the solar system.

What We Learned on Mars[*]

By Ivan Semeniuk
New Scientist, January 7, 2006

It was supposed to be a 100-day mission to a planet that eats spacecraft for breakfast. But NASA's twin rovers, Spirit and Opportunity, have now spent two years toddling around the frigid wasteland that is Mars. They have survived more than seven times as long and driven 10 times as far as they were designed to. To match this feat, the Hubble Space Telescope would need to remain in orbit for another 55 years.

The rovers are not just successful, they are outrageously so. With the next rover mission already under development, it is worth asking why Spirit and Opportunity have led such a charmed existence.

The trouble with success is it's hard to pin down. When things go wrong in space the reasons are not always immediately obvious, but in most cases they can be isolated. A bad bit of software, for example, led to the explosion of the Ariane 5 rocket on its maiden flight in 1996, and a failure to appreciate what a piece of foam can do when travelling at high speed doomed the space shuttle Columbia in 2003.

When things go right the reasons are more diffuse. Every component of a successful mission lays claim to some part of the overall achievement. The catch is that while some are crucial to success, others are benignly neutral, and still others harmful, but not enough to compromise the mission. Then there is luck.

The rovers have been lucky in one very big way: the wind has been with them. At the outset, engineers imagined that even if Spirit and Opportunity landed safely and operated correctly, a build-up of dust on their solar panels would fairly quickly lead to their demise. The rate of dust deposition on Mars was measured during the Pathfinder mission in 1997 and was factored in to the rovers' expected lifespan.

What engineers could not have predicted from the Pathfinder data is that random gusts of wind can sometimes sweep solar panels clean. Both rovers have had their lives extended by "cleaning events" of this kind.

The unexpected benefit of wind merely shifts the most likely cause of the rovers' mortality to mechanical failure. Yet they have not failed. This is almost certainly because every piece of hardware was tested to three times the expected lifetime of the mission with no recorded failures.

The testing regime suggests that the real reason the rovers have managed so well on Mars has to do with what happened before they left Earth. According to Steven Squyres, the mission's principal investigator, the rovers "arose out of catastrophe", namely the double failure of the Mars Climate Orbiter and the Mars Polar Lander in 1999, and NASA's subsequent shakedown of its entire Mars programme. As Pete Theisinger, deputy director for Mars exploration at NASA's Jet Propulsion Laboratory, puts it: "People were very paranoid" about the prospect of a hidden flaw lurking in the hardware or design of the rover mission. This made team members more likely to identify potential problems and "work them to death". It also led to conservative choices when it came to determining what the rovers should do. Instead of doing more, they were built to do less—but more reliably.

Concern over a negative outcome led to improved communication among team members and, in particular, between engineers and scientists. On most missions the interaction between these two camps inevitably leads to a certain tension over spacecraft functionality versus scientific return. During the rover mission, perhaps as never before, this traditional divide dissolved. "If you had walked in on one of our meetings, you would have had a hard time telling the scientists from the engineers," says John Callas, the deputy project manager.

This unusual harmony is exemplified by Squyres himself, who has earned a reputation as a scientist who doesn't mind getting his hands dirty. He made a point of knowing every nut and bolt on Spirit and Opportunity and involved himself at every stage of development. By all accounts, his strong, charismatic leadership had a direct impact on testing. Engineers learned to trust that when Squyres was asking for something to be done it must be important. The bottom line is that even on uncrewed missions, these kinds of human factors are the key to success.

In the light of this it may seem strange that NASA has chosen an alternative route for the next rover mission to Mars. When the Mars Science Laboratory launches in 2009 it will not have a single principal investigator in a central leadership role and its rover will use a different method of landing from the one that was so thoroughly tested for Spirit and Opportunity. The reasons have to do with the size and complexity of the new rover, which will have capabilities far beyond the existing pair.

Managers at NASA are hoping these things may not matter, given what they have learned from the success of Spirit and Opportunity. The assumption is that key elements—such as team chemistry—can be reproduced in a different context. Let's hope that's right, otherwise the planet that eats spacecraft for breakfast could well have another on its plate.

Lander Finds Ice on Mars, Scientists Say[*]

By David Brown
The Washington Post, June 21, 2008

Scientists with the Phoenix Mars mission yesterday declared for certain that there is ice on the Red Planet, putting them an essential step closer to answering the question that has driven three decades of Mars exploration and centuries of Earth-bound speculation: Could there have been life there?

Pictures beamed 170 million miles to Earth from the Phoenix lander atop Mars's northern polar plain erased any doubt about the presence of ice, they said.

But the evidence came in a roundabout way. Last Sunday, several dice-size solids were observed at the bottom of a trench that had been dug by Phoenix's robotic arm. On Thursday, they were gone.

The only reasonable explanation, the scientists said, is that the objects were pieces of ice that evaporated into the dry Martian atmosphere through a process called sublimation. And the presence of ice means that Mars might once have had liquid water, which is essential for life—at least as it is known on Earth.

It is too soon to know whether the entire astrophysical community will accept the disappearing objects reported yesterday as proof, but the Phoenix researchers said they do not need any more convincing.

The rocket thrusters that slowed Phoenix to a soft landing revealed a white, hard substance in the ground beneath it—and tantalizingly out of reach—when the lander touched down on May 25. Similar white material was visible when the robotic arm began to dig below the top few inches of Martian soil.

One possibility was that it was salt of some sort. But ice was always the more likely explanation.

"Salt does not behave like that," said Mark Lemmon, a scientist at Texas A&M University who is in charge of Phoenix's stereo surface imager. "We found what we were looking for. This tells us we have water ice within reach of the arm."

Although Mars is much too cold now to have liquid water on its surface, scientists believe that may not have always been the case. Images from as far back as the Viking missions in the 1970s revealed channels and gullies that appear to have been carved by flowing liquid at some point in the planet's history.

The Mars Odyssey orbiter, using a device called a gamma ray spectrometer, proved in 2002 that huge quantities of hydrogen existed under the Martian topsoil. Although many compounds are high in hydrogen (including petroleum), the scientists believe the only one that could be there in such quantity is water ice, which consists of two hydrogen atoms bonded to an oxygen atom.

"I don't know how you could have so much hydrogen under the surface, and something that disappears at just the temperature of ice, and have it not be ice," said Peter Smith, a physicist at the University of Arizona who is the principal investigator for the Phoenix mission.

The researchers chose Mars's northern polar plain as the craft's landing site specifically because they believed it would give them the best chance of finding ice. They now believe their hunch was correct.

"If you were to get a big broom and sweep it off, we are on a big ice sheet," Smith said.

Water is necessary for life but is far from sufficient. Scientists will also have to find a fair amount of carbon before they are willing to say the planet might have been habitable. Carbon forms the chemical backbone of proteins and fats, and in addition to water it is the major constituent of living cells and tissue.

"The ice may always be in a frozen state," Smith said, noting that without liquid water, the formation of life is hard to imagine. "If you have ice and no food, it is not a habitable zone."

The researchers are extremely interested to know what compounds may be frozen into the ice or, more likely, contained in the dry, reddish soil covering it.

Among the instruments on the lander is a panel of miniature ovens in which Martian soil will be heated to about 1,800 degrees Fahrenheit. Compounds containing carbon will be vaporized and then chemically analyzed.

The first load of dirt is cooking and will probably be finished today, Lemmon said. Analyzing the results will then take several days.

The researchers hope to get a piece of ice into one of the ovens at some point. Once sealed off from the atmosphere and heated, it would probably have a brief life as liquid water before becoming vapor, Lemmon said.

If the scientists find carbon compounds of the right size and makeup, they will be able to refine their guesses about Mars's life-supporting potential. However, no instrument on the Phoenix mission will be able to prove with certainty that life once existed there—or that it never has.

Life On Mars?[*]

By Carl Zimmer
Smithsonian, May 2005

On August 7, 1996, reporters, photographers and television camera operators surged into NASA headquarters in Washington, D.C. The crowd focused not on the row of seated scientists in NASA's auditorium but on a small, clear plastic box on the table in front of them. Inside the box was a velvet pillow, and nestled on it like a crown jewel was a rock—from Mars. The scientists announced that they'd found signs of life inside the meteorite. NASA administrator Daniel Goldin gleefully said it was an "unbelievable" day. He was more accurate than he knew.

The rock, the researchers explained, had formed 4.5 billion years ago on Mars, where it remained until 16 million years ago, when it was launched into space, probably by the impact of an asteroid. The rock wandered the inner solar system until 13,000 years ago, when it fell to Antarctica. It sat on the ice near Allan Hills until 1984, when snowmobiling geologists scooped it up.

Scientists headed by David McKay of the Johnson Space Center in Houston found that the rock, called ALH84001, had a peculiar chemical makeup. It contained a combination of minerals and carbon compounds that on Earth are created by microbes. It also had crystals of magnetic iron oxide, called magnetite, which some bacteria produce. Moreover, McKay presented to the crowd an electron microscope view of the rock showing chains of globules that bore a striking resemblance to chains that some bacteria form on Earth. "We believe that these are indeed microfossils from Mars," McKay said, adding that the evidence wasn't "absolute proof" of past Martian life but rather "pointers in that direction."

Among the last to speak that day was J. William Schopf, a University of California at Los Angeles paleobiologist, who specializes in early Earth fossils. "I'll show you the oldest evidence of life on this planet," Schopf said to the audience, and displayed a slide of a 3.465 billion-year-old fossilized chain of microscopic globules that he had found in Australia. "These are demonstrably fossils," Schopf said, implying that NASA's Martian pictures were not. He closed by quoting the astronomer Carl Sagan: "Extraordinary claims require extraordinary evidence."

Despite Schopf's note of skepticism, the NASA announcement was trumpeted worldwide. "Mars lived, rock shows Meteorite holds evidence of life on another world," said the *New York Times*. "Fossil from the red planet may prove that we are not alone," declared *The Independent of London*.

Over the past nine years, scientists have taken Sagan's words very much to heart. They've scrutinized the Martian meteorite (which is now on view at the Smithsonian's National Museum of Natural History), and today few believe that it harbored Martian microbes.

The controversy has prompted scientists to ask how they can know whether some blob, crystal or chemical oddity is a sign of life—even on Earth. A debate has flared up over some of the oldest evidence for life on Earth, including the fossils that Schopf proudly displayed in 1996. Major questions are at stake in this debate, including how life first evolved on Earth. Some scientists propose that for the first few hundred million years that life existed, it bore little resemblance to life as we know it today.

NASA researchers are taking lessons from the debate about life on Earth to Mars. If all goes as planned, a new generation of rovers will arrive on Mars within the next decade. These missions will incorporate cutting-edge biotechnology designed to detect individual molecules made by Martian organisms, either living or long dead.

The search for life on Mars has become more urgent thanks in part to probes by the two rovers now roaming Mars' surface and another spaceship that is orbiting the planet. In recent months, they've made a series of astonishing discoveries that, once again, tempt scientists to believe that Mars harbors life—or did so in the past. At a February conference in the Netherlands, an audience of Mars experts was surveyed about Martian life. Some 75 percent of the scientists said they thought life once existed there, and of them, 25 percent think that Mars harbors life today.

The search for the fossil remains of primitive single-celled organisms like bacteria took off in 1953, when Stanley Tyler, an economic geologist at the University of Wisconsin, puzzled over some 2.1 billion-year-old rocks he'd gathered in Ontario, Canada. His glassy black rocks known as cherts were loaded with strange, microscopic filaments and hollow balls. Working with Harvard paleobotanist Elso Barghoorn, Tyler proposed that the shapes were actually fossils, left behind by ancient life-forms such as algae. Before Tyler and Barghoorn's work, few fossils had been found that predated the Cambrian Period, which began about 540 million years ago. Now the two scientists were positing that life was present much earlier in the 4.55 billion-year history of our planet. How much further back it went remained for later scientists to discover.

In the next decades, paleontologists in Africa found 3 billion-year-old fossil traces of microscopic bacteria that had lived in massive marine reefs. Bacteria can also form what are called biofilms, colonies that grow in thin layers over surfaces such as rocks and the ocean floor, and scientists have found solid evidence for biofilms dating back 3.2 billion years.

But at the time of the NASA press conference, the oldest fossil claim belonged to UCLA's William Schopf, the man who spoke skeptically about NASA's finds at the same conference. During the 1960s, '70s and '80s, Schopf had become a leading expert on early life-forms, discovering fossils around the world, including 3 billion-year-old fossilized bacteria in South Africa. Then, in 1987, he and some colleagues reported that they had found the 3.465 billion-year-old microscopic fossils at a site called Warrawoona in the Western Australia outback—the ones he would display at the NASA press conference. The bacteria in the fossils were so sophisticated, Schopf says, that they indicate "life was flourishing at that time, and thus, life originated appreciably earlier than 3.5 billion years ago."

Since then, scientists have developed other methods for detecting signs of early life on Earth. One involves measuring different isotopes, or atomic forms, of carbon; the ratio of the isotopes indicates that the carbon was once part of a living thing. In 1996, a team of researchers reported that they had found life's signature in rocks from Greenland dating back 3.83 billion years.

The signs of life in Australia and Greenland were remarkably old, especially considering that life probably could not have persisted on Earth for the planet's first few hundreds of millions of years. That's because asteroids were bombarding it, boiling the oceans and likely sterilizing the planet's surface before about 3.8 billion years ago. The fossil evidence suggested that life emerged soon after our world cooled down. As Schopf wrote in his book *Cradle of Life*, his 1987 discovery "tells us that early evolution proceeded very far very fast."

A quick start to life on Earth could mean that life could also emerge quickly on other worlds—either Earth-like planets circling other stars, or perhaps even other planets or moons in our own solar system. Of these, Mars has long looked the most promising.

The surface of Mars today doesn't seem like the sort of place hospitable to life. It is dry and cold, plunging down as far as -220 degrees Fahrenheit. Its thin atmosphere cannot block ultraviolet radiation from space, which would devastate any known living thing on the surface of the planet. But Mars, which is as old as Earth, might have been more hospitable in the past. The gullies and dry lake beds that mark the planet indicate that water once flowed there. There's also reason to believe, astronomers say, that Mars' early atmosphere was rich enough in heat-trapping carbon dioxide to create a greenhouse effect, warming the surface. In other words, early Mars was a lot like early Earth. If Mars had been warm and wet for millions or even billions of years, life might have had enough time to emerge. When conditions on the surface of Mars turned nasty, life may have become extinct there. But fossils may have been left behind. It's even possible that life could have survived on Mars below the surface, judging from some microbes on Earth that thrive miles underground.

When NASA's McKay presented his pictures of Martian fossils to the press that day in 1996, one of the millions of people who saw them on television was a young British environmental microbiologist named Andrew Steele. He had just earned a PhD at the University of Portsmouth, where he was studying bacterial biofilms that can absorb radioactivity from contaminated steel in nuclear facilities. An expert at microscopic images of microbes, Steele got McKay's telephone number from directory assistance and called him. "I can get you a better picture than that," he said, and convinced McKay to send him pieces of the meteorite. Steele's analyses were so good that soon he was working for NASA.

Ironically, though, his work undercut NASA's evidence: Steele discovered that Earthly bacteria had contaminated the Mars meteorite. Biofilms had formed and spread through cracks into its interior. Steele's results didn't disprove the Martian fossils outright—it's possible that the meteorite contains both Martian fossils and Antarctic contaminants—but, he says, "The problem is, how do you tell the difference?" At the same time, other scientists pointed out that nonliving processes on Mars also could have created the globules and magnetite clumps that NASA scientists had held up as fossil evidence.

But McKay stands by the hypothesis that his microfossils are from Mars, saying it is "consistent as a package with a possible biological origin." Any alternative explanation must account for all of the evidence, he says, not just one piece at a time.

The controversy has raised a profound question in the minds of many scientists: What does it take to prove the presence of life billions of years ago? In 2000, Oxford paleontologist Martin Brasier borrowed the original Warrawoona fossils from the Natural History Museum in London, and he and Steele and their colleagues have studied the chemistry and structure of the rocks. In 2002, they concluded that it was impossible to say whether the fossils were real, essentially subjecting Schopf's work to the same skepticism that Schopf had expressed about the fossils from Mars. "The irony was not lost on me," says Steele.

In particular, Schopf had proposed that his fossils were photosynthetic bacteria that captured sunlight in a shallow lagoon. But Brasier and Steele and co-workers concluded that the rocks had formed in hot water loaded with metals, perhaps around a superheated vent at the bottom of the ocean—hardly the sort of place where a sun-loving microbe could thrive. And microscopic analysis of the rock, Steele says, was ambiguous, as he demonstrated one day in his lab by popping a slide from the Warrawoona chert under a microscope rigged to his computer. "What are we looking at there?" he asks, picking a squiggle at random on his screen. "Some ancient dirt that's been caught in a rock? Are we looking at life? Maybe, maybe. You can see how easily you can fool yourself. There's nothing to say that bacteria can't live in this, but there's nothing to say that you are looking at bacteria."

Schopf has responded to Steele's criticism with new research of his own. Analyzing his samples further, he found that they were made of a form of carbon known as kerogen, which would be expected in the remains of bacteria. Of his critics, Schopf says, "they would like to keep the debate alive, but the evidence is overwhelming."

The disagreement is typical of the fast-moving field. Geologist Christopher

Fedo of George Washington University and geochronologist Martin Whitehouse of the Swedish Museum of Natural History have challenged the 3.83 billion-year-old molecular trace of light carbon from Greenland, saying the rock had formed from volcanic lava, which is much too hot for microbes to withstand. Other recent claims also are under assault. A year ago, a team of scientists made headlines with their report of tiny tunnels in 3.5 billion-year-old African rocks. The scientists argued that the tunnels were made by ancient bacteria around the time the rock formed. But Steele points out that bacteria might have dug those tunnels billions of years later. "If you dated the London Underground that way," says Steele, "you'd say it was 50 million years old, because that's how old the rocks are around it."

Such debates may seem indecorous, but most scientists are happy to see them unfold. "What this will do is get a lot of people to roll up their sleeves and look for more stuff," says MIT geologist John Grotzinger. To be sure, the debates are about subtleties in the fossil record, not about the existence of microbes long, long ago. Even a skeptic like Steele remains fairly confident that microbial biofilms lived 3.2 billion years ago. "You can't miss them," Steele says of their distinctive weblike filaments visible under a microscope. And not even critics have challenged the latest from Minik Rosing, of the University of Copenhagen's Geological Museum, who has found the carbon isotope life signature in a sample of 3.7 billion-year-old rock from Greenland—the oldest undisputed evidence of life on Earth.

At stake in these debates is not just the timing of life's early evolution, but the path it took. This past September, for example, Michael Tice and Donald Lowe of Stanford University reported on 3.416 billion-year-old mats of microbes preserved in rocks from South Africa. The microbes, they say, carried out photosynthesis but didn't produce oxygen in the process. A small number of bacterial species today do the same—anoxygenic photosynthesis, it's called—and Tice and Lowe suggest that such microbes, rather than the conventionally photosynthetic ones studied by Schopf and others, flourished during the early evolution of life. Figuring out life's early chapters will tell scientists not only a great deal about the history of our planet. It will also guide their search for signs of life elsewhere in the universe—starting with Mars.

In January 2004, the NASA rovers Spirit and Opportunity began rolling across the Martian landscape. Within a few weeks, Opportunity had found the best evidence yet that water once flowed on the planet's surface. The chemistry of rock it sampled from a plain called Meridiani Planum indicated that it had formed billions of years ago in a shallow, long-vanished sea. One of the most important results of the rover mission, says Grotzinger, a member of the rover science team, was the robot's observation that rocks on Meridiani Planum don't seem to have been crushed or cooked to the degree that Earth rocks of the same age have been—their crystal structure and layering remain intact. A paleontologist couldn't ask for a better place to preserve a fossil for billions of years.

The past year has brought a flurry of tantalizing reports. An orbiting probe and ground-based telescopes detected methane in the atmosphere of Mars. On Earth, microbes produce copious amounts of methane, although it can also be produced by volcanic activity or chemical reactions in the planet's crust. In February, reports raced through the media about a NASA study allegedly concluding that the Martian methane might have been produced by underground microbes. NASA headquarters quickly swooped in—perhaps worried about a repeat of the media frenzy surrounding the Martian meteorite—and declared that it had no direct data supporting claims for life on Mars.

But just a few days later, European scientists announced that they had detected formaldehyde in the Martian atmosphere, another compound that, on Earth, is produced by living things. Shortly thereafter, researchers at the European Space Agency released images of the Elysium Plains, a region along Mars' equator. The texture of the landscape, they argued, shows that the area was a frozen ocean just a few million years ago—not long, in geological time. A frozen sea may still be there today, buried under a layer of volcanic dust. While water has yet to be found on Mars' surface, some researchers studying Martian gullies say that the features may have been produced by underground aquifers, suggesting that water, and the life-forms that require water, might be hidden below the surface.

Andrew Steele is one of the scientists designing the next generation of equipment to probe for life on Mars. One tool he plans to export to Mars is called a microarray, a glass slide onto which different antibodies are attached. Each antibody recognizes and latches onto a specific molecule, and each dot of a particular antibody has been rigged to glow when it finds its molecular partner. Steele has preliminary evidence that the microarray can recognize fossil hopanes, molecules found in the cell walls of bacteria, in the remains of a 25 million-year-old biofilm.

This past September, Steele and his colleagues traveled to the rugged Arctic island of Svalbard, where they tested the tool in the area's extreme environment as a prelude to deploying it on Mars. As armed Norwegian guards kept a lookout for polar bears, the scientists spent hours sitting on chilly rocks, analyzing fragments of stone. The trip was a success: the microarray antibodies detected proteins made by hardy bacteria in the rock samples, and the scientists avoided becoming food for the bears.

Steele is also working on a device called MASSE (Modular Assays for Solar System Exploration), which is tentatively slated to fly on a 2011 European Space Agency expedition to Mars. He envisions the rover crushing rocks into powder, which can be placed into MASSE, which will analyze the molecules with a microarray, searching for biological molecules.

Sooner, in 2009, NASA will launch the Mars Science Laboratory Rover. It's designed to inspect the surface of rocks for peculiar textures left by biofilms. The Mars lab may also look for amino acids, the building blocks of proteins, or other organic compounds. Finding such compounds wouldn't prove the existence of life on Mars, but it would bolster the case for it and spur NASA scientists to look more closely.

Difficult as the Mars analyses will be, they're made even more complex by the threat of contamination. Mars has been visited by nine spacecraft, from Mars 2, a Soviet probe that crashed into the planet in 1971, to NASA's Opportunity and Spirit. Any one of them might have carried hitchhiking Earth microbes. "It might be that they crash-landed and liked it there, and then the wind could blow them all over the place," says Jan Toporski, a geologist at the University of Kiel, in Germany. And the same interplanetary game of bumper cars that hurtled a piece of Mars to Earth might have showered pieces of Earth on Mars. If one of those terrestrial rocks was contaminated with microbes, the organisms might have survived on Mars—for a time, at least—and left traces in the geology there. Still, scientists are confident they can develop tools to distinguish between imported Earth microbes and Martian ones.

Finding signs of life on Mars is by no means the only goal. "If you find a habitable environment and don't find it inhabited, then that tells you something," says Steele. "If there is no life, then why is there no life? The answer leads to more questions." The first would be what makes life-abounding Earth so special. In the end, the effort being poured into detecting primitive life on Mars may prove its greatest worth right here at home.

SOL 20　　　　　SOL 24

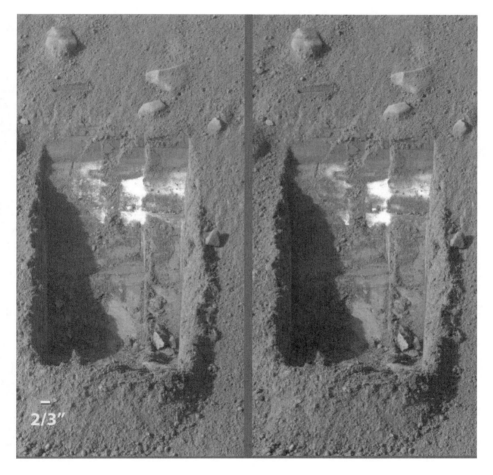

2/3"

Images of sublimation of ice taken by NASA's Phoenix Mars Lander's Surface Stereo Imager on the 21st and 25th days of the mission, or Sols 20 and 24 (June 15 and 19, 2008).

5

Sailing Through the Solar System:
A Look at Unmanned Probes

Photo of Saturn taken by the Cassini Orbiter.

Editor's Introduction

Even as the debate about manned spaceflight's purpose—for sightseeing or science?—and destination—to return to the Moon or go straight on to Mars?—continues unabated, robotic space probes are at this moment patrolling our solar system and gathering valuable scientific data. To some they are the advance scouts for future manned missions; to others, they are expert explorers in their own right, fully capable of performing full-up missions. Given the current level of technology, they are also quite necessary. The environment on planets like Venus is far too hostile for human beings, and the outer gas giants—Jupiter, Saturn, Uranus, and Neptune—are simply too distant. However the next age of space exploration unfolds, it is clear that unmanned probes will have a vital role in it.

In the broadest sense, an unmanned space probe is any robotic spacecraft that leaves the Earth's gravity on an exploratory mission. Such probes have been a central component of space exploration since the 1950s. Though *Sputnik 1*, launched in 1957, was the first artificial satellite, it was not the first space probe because it never left the gravity well of the Earth to move out into space. That distinction goes to another Soviet probe, *Luna 1*, which conducted the first flyby of the Moon in 1959. Since that time unmanned probes have grown far more sophisticated, moving from relatively simple flybys of celestial bodies to orbiters, which are captured by the gravity of a moon or planet. This advancement came with the development of onboard guidance systems capable of making needed course adjustments; the early flybys had no such systems. Today, probes can be directed from Earth via radio, although some of the most sophisticated modern probes, such as the *Cassini–Huygens* spacecraft now exploring Saturn and its moons, have software that allows them to act with a high degree of autonomy for extended periods.

Techniques for landing probes on celestial bodies have also improved dramatically, moving from early "impactor" missions, in which spacecraft literally crashed onto the surface of their targets, to lander missions in which a spacecraft is gently brought to rest on a celestial body's surface via parachutes, retrorockets, or airbags. Upon reaching the surface, some landers have released rovers. Though rover missions are limited in range (at least as compared to how much ground a human being could cover), they have given Earth-bound mission controllers something stationary landers could not—the power to tour an alien landscape by remote control.

Because Martian probes were covered extensively in the previous section, the focus of this fifth chapter will be on probes that are touring the rest of the solar system and beyond. In addition to detailing the missions of some recent probes, the collected articles will also provide an overview of famous ones, such as *Pioneer 10*, *Pioneer 11*, *Voyager 1*, and *Voyager 2*—all of which have now become "interstellar," since passing through the outer solar system's heliopause. Though launched in the 1970s, both Voyager probes are still transmitting data back to Earth. Since completing its mission to Jupiter and Saturn, *Voyager 1* has become the furthest manmade object from Earth. At its present speed, it will reach the star AC+79 3888 in the constellation of Ophiuchus in about 40,000 years. (Much before that, *New Horizons*, the American spacecraft launched to Pluto in 2006, will join this quartet of probes in interstellar space after it completes its mission to the dwarf planet in 2015.)

In this section's first article, "Ask the Experts," Jeremy Jones, chief of the navigation team for the Cassini Project at NASA's Jet Propulsion Laboratories, provides a vivid description of how unmanned probes accurately navigate space to reach their objectives. In "Spacecraft Behaving Badly," Neil deGrasse Tyson, director of the Hayden Planetarium at the American Museum of Natural History, describes the phenomenon research scientists are now calling the "Pioneer anomaly"—a mismatch between calculations predicting where the two spacecraft should be as opposed to where they are. This mismatch may help to provide new insights into the physical laws governing the universe. Tim Appenzeller, in "Go Boldly, Voyager," celebrates *Voyager 1* and *Voyager 2* as they move into interstellar space. "NASA in the 20th century is going to be remembered for two things: Apollo moon landings and Voyager," says one scientist in Appenzeller's article. Finally, Dan Cray provides a report on the 11 interplanetary probes currently moving about the solar system in "Cosmic Flock."

Ask the Experts[*]

Scientific American, March 27, 2006

How do space probes navigate large distances with such accuracy and how do the mission controllers know when they've reached their target?

Jeremy Jones, chief of the navigation team for the Cassini Project at NASA's Jet Propulsion Laboratory, offers this explanation.

The accurate navigation of space probes depends on four factors: First is the measurement system for determining the position and speed of a probe. Second is the location from which the measurements are taken. Third is an accurate model of the solar system, and fourth, models of the motion of a probe.

For all U.S. interplanetary probes, the antennas of the Deep Space Network (DSN) act as the measurement system. These antennas transmit radio signals to a probe, which receives these signals and, with a slight frequency shift, returns them to the ground station. By computing the difference between the transmitted and received signals, a probe's distance and speed along the line from the antenna can be determined with great accuracy, thanks to the high frequency of the signals and a very accurate atomic clock by which to measure the small frequency changes. By combining these elements, navigators can measure a probe's instantaneous line-of-sight velocity and range to an accuracy of 0.05 millimeter-per-second and three meters respectively, relative to the antenna.

Many probes also carry cameras that are used to image the destination, whether it be a moon, planet or other body. During the final approach, these images are used when the distance becomes small. For example, the Cassini spacecraft's camera provides an angular measurement with an accuracy of three microradians (three kilometers) at a distance of one million kilometers. The images complement the radio data and provide a direct tie to the target.

Calculation of the trajectory of a space probe requires the use of an inertial coordinate system as well, wherein a grid is laid over the solar system and fixed relative to the star background. For interplanetary missions, an inertial coordinate system with an origin at the center of mass of the solar system is used. Because the measurements provide information on the position of a probe relative to the antenna, knowledge of the antenna's location relative to this inertial coordinate system is used to convert the measurements into elements in the system. Where the antenna is depends not only on its geographic location on Earth's surface, but on Earth's position relative to the solar system center of mass (known as the Earth ephemeris). Measurements of this ephemeris have an accuracy of about 0.5 kilometer and the location of the antenna is known to an accuracy of better than five centimeters.

The third component of interplanetary navigation is an accurate model of the solar system. Gravity is the most important force acting on a spacecraft. Determining these gravitational forces requires accurate knowledge of the locations of all of the major bodies, such as the sun and all the planets, over the course of time. This information is provided by the planetary ephemeris, which has been in continuous development since the beginning of the interplanetary space program. Thanks to this longstanding work, the location of Saturn was known to an accuracy of a few hundreds of kilometers previous to Cassini's final approach when it deployed its camera. After the probe entered Saturn orbit, the moons of the giant planet became important gravitational bodies. Their locations have been determined to an accuracy of a few kilometers relative to Saturn.

The final component of accurate navigation takes all of these other elements and, using models of the forces acting on a probe and orbital dynamics, estimates its location. By taking regular measurements over a period of time, a probe's position and velocity can be determined. For example, Cassini's location is typically determined to a kilometer or less relative to Saturn. Using a probe's known position and velocity, its future positions can be worked out. Navigators compare these positions to the predicted location of the target body—based on the ephemeris—to determine when a probe will reach its target. Then, all that's left to do is to collect the flyby data, take a deep breath, and go on to the next encounter.

Spacecraft Behaving Badly[*]

By Neil deGrasse Tyson
Natural History, April 2008

There's no sweeping it under the rug. NASA's twin Pioneer 10 and Pioneer 11 space probes, launched in the early 1970s and headed for stars in the depths of our galaxy, are both experiencing a mysterious, continuous force that is altering their expected trajectories. Calculations say the Pioneers should each be in a particular place, but the probes themselves have told us they're each someplace else—as much as a quarter million miles closer to the Sun than they're supposed to be.

That mismatch, known as the Pioneer anomaly, first became evident in the early 1980s, by which time the spacecraft were so far from the Sun that the slight outward pressure of sunlight no longer exerted significant influence over their velocity. Scientists expected that Newtonian gravity alone would thenceforth account for the pace of the Pioneers' journey. But things seemingly haven't turned out that way. The extra little push from solar radiation had been masking an anomaly. Once the Pioneers reached the point where the sunlight's influence was less than the anomaly's, both spacecraft began to register an unexplained, persistent change in velocity—a sunward force, a drag—operating at the rate of a couple hundred-millionths of an inch per second for every second of time the twins have been traveling. That may not sound like much, but it eventually claimed thousands of miles of lost ground for every year out on the road.

Contrary to stereotype, research scientists don't sit around their offices smugly celebrating their mastery of cosmic truths. Nor are scientific discoveries normally heralded by people in lab coats proclaiming "Eureka!" Instead, researchers say things like "Hmm, that's odd." From such humble beginnings come mostly dead ends and frustration, but also an occasional new insight into the laws of the universe.

Once the Pioneer anomaly revealed itself, scientists said, "Hmm, that's odd." So they kept looking, and the oddness didn't go away. Serious investigation began in 1994, the first research paper about it appeared in 1998, and since then all sorts of explanations have been proffered to account for the anomaly. Contenders that have now been ruled out include software bugs, leaky valves in the midcourse-correction rockets, the solar wind interacting with the probes' radio signals, the probes' magnetic fields interacting with the Sun's magnetic field, the gravity exerted by newly discovered Kuiper Belt objects, the deformability of space and time, and the accelerating expansion of the universe. The remaining explanations range from the everyday to the exotic. Among them is the suspicion that in the outer solar system, Newtonian gravity begins to fail.

The very first spacecraft in the Pioneer program—Pioneer 0 (that's right, "zero")—was launched, unsuccessfully, in the summer of 1958. Fourteen more were launched over the next two decades. Pioneers 3 and 4 studied the Moon; 5 through 9 monitored the Sun; 10 flew by Jupiter; 11 flew by Jupiter and Saturn; 12 and 13 visited Venus.

Pioneer 10 left Cape Canaveral on the evening of March 2, 1972—nine months before the Apollo program's final Moon landing—and crossed the Moon's orbit the very next morning. In July 1972 it became the first human-made object to traverse the asteroid belt, the band of rocky rubble that separates the inner solar system from the giant outer planets. In December 1973 it became the first to get a "gravity assist" from massive Jupiter, which helped kick it out of the solar system for good. Although NASA planned for Pioneer 10 to keep signaling Earth for a mere twenty-one months, the craft's power sources kept going and going—enabling the fellow to call home for thirty years, until January 22, 2003. Its twin, Pioneer II, had a shorter signaling life, with its final transmission arriving on September 30, 1995.

At the heart of Pioneers 10 and 11 is a toolbox-size equipment compartment, from which booms holding instruments and a miniature power plant project at various angles. More instruments and several antennas are clamped to the compartment itself: Heat-responsive louvers keep the onboard electronics at ideal operating temperatures, and there are three pairs of rocket thrusters, packed with reliable propellant, to help with alignments and midcourse corrections en route to Jupiter.

Power for the twins and their fifteen scientific instruments comes from radioactive chunks of plutonium-238, which drive four radioisotope thermoelectric generators, sensibly abbreviated RTGs. The heat from the slowly decaying plutonium, with its half-life of eighty-eight years, yielded enough electricity to run the spacecraft, photograph Jupiter and its satellites in multiple wavelengths, record sundry cosmic phenomena, and conduct experiments more or less continuously for upwards of a decade. But by April 2001 the signal from Pioneer 10 had dwindled to a barely detectable billionth of a trillionth of a watt.

The probes' main agent of communication is a nine-foot-wide dish-shaped antenna pointed toward Earth. To preserve the antenna's alignment, each spacecraft has star and Sun sensors that keep it spinning along the antenna's central axis in much the way that a quarterback spins a football around its long axis to stabilize the ball's trajectory. For the duration of the dish antenna's prolonged life, it sent and received radio signals via the Deep Space Network, an ensemble of sensitive antennas that span the globe, making it possible for engineers to monitor the spacecraft without a moment's interruption.

The famous finishing touch on Pioneers 10 and 11 is a gold-plated plaque affixed to the side of the craft. The plaque includes an engraved illustration of a naked adult male and female; a sketch of the spacecraft itself; shown in correct proportion to the humans; and a diagram of the Sun's position in the Milky Way, announcing the spacecraft's provenance to any intelligent aliens who might stumble across one of the twins. (I've always had my doubts about this cosmic calling card. Most people wouldn't give their home address to a stranger in the street, even when the stranger is one of our own species. Why, then, give our home address to aliens from another planet?)

Space travel involves a lot of coasting. Typically, a spacecraft relies on rockets to get itself off the ground and on its way. Other, smaller engines may fire en route to refine the craft's trajectory or pull the craft into orbit around a target object. In between, it simply coasts. For engineers to calculate a craft's Newtonian trajectory between any two points in the solar system, they must account for every single source of gravity along the way, including comets, asteroids, moons, and planets. As an added challenge, they must aim for where the target should be when the spacecraft is due to arrive, not for the target's current location.

Calculations completed, off went Pioneers 10 and 11 on their multibillion-mile journeys through interplanetary space—boldly going where no hardware had gone before, and opening new vistas on the planets of our solar system. Little did anyone foresee that in their twilight years the twins would also become unwitting probes of the fundamental laws of gravitational physics.

Astrophysicists do not normally discover new laws of nature. We cannot manipulate the objects of our scrutiny. Our telescopes are passive probes that cannot tell the cosmos what to do. Yet they can tell us when something isn't following orders. Take the planet Uranus, whose discovery is credited to the English astronomer William Herschel and dated to 1781 (others had already noted its presence in the sky but misidentified it as a star). As observational data about its orbit accumulated over the following decades, people began to notice that Uranus deviated slightly from the dictates of Newton's laws of gravity, which by then had withstood a century's worth of testing on the other planets and their moons. Some prominent astronomers suggested that perhaps Newton's laws begin to break down at such great distances from the Sun.

What to do? Abandon or modify Newton's laws and dream up new rules of gravity? Or postulate a yet-to-be-discovered planet in the outer solar system, whose gravity was absent from the calculations for Uranus's orbit? The answer came in 1846, when astronomers discovered the planet Neptune just where a planet had to be for its gravity to perturb Uranus in just the ways measured. Newton's laws were safe . . . for the time being.

Then there's Mercury, the planet closest to the Sun. Its orbit, too, habitually disobeyed Newton's laws of gravity. Having predicted Neptune's position on the sky within one degree, the French astronomer Urbain-Jean-Joseph Le Verrier now postulated two possible causes for Mercury's deviant behavior. Either it was another new planet (call it Vulcan) orbiting so close to the Sun that it would be well-nigh impossible to discover in the solar glare, or it was an entire, uncataloged belt of asteroids orbiting between Mercury and the Sun.

Turns out Le Verrier was wrong on both counts. This time he really did need a new understanding of gravity. Within the limits of precision that our measuring tools impose, Newton's laws behave well in the outer solar system. However, they break down in the inner solar system, where they are superseded by Einstein's general relativity. The closer you are to the Sun, the less you can ignore the exotic effects of its powerful gravitational field.

TWO PLANETS. TWO SIMILAR-LOOKING ANOMALIES. TWO
COMPLETELY DIFFERENT EXPLANATIONS.

Pioneer 10 had been coasting through space for less than a decade and was around 15 AU from the Sun when John D. Anderson, a specialist in celestial mechanics and radio-wave physics at NASA's Jet Propulsion Laboratory (JPL), first noticed that the data were drifting away from the predictions made by JPL's computer model. (One AU, or astronomical refit, represents the average distance between Earth and the Sun; it's a "yardstick" for measuring distances within the solar system.) By the time Pioneer 10 reached 20 AU, a distance at which pressure from the Sun's rays no longer mattered much to the trajectory of the spacecraft, the drift was unmistakable. Initially Anderson didn't fuss over the discrepancy, thinking the problem could probably be blamed on either the software or the spacecraft itself. But he soon determined that only if he added to the equations an invented force—a constant change in velocity (an acceleration) back toward the Sun for every second of the trip—would the location predicted for Pioneer 10's signal match the location of its actual signal.

Had Pioneer 10 encountered something unusual along its path? If so, that could explain everything. Nope. Pioneer 11 was heading out of the solar system in a whole other direction, yet it, too, required an adjustment to its predicted location. In fact, Pioneer 11's anomaly is somewhat larger than Pioneer 10's.

Faced with either revising the tenets of conventional physics or seeking ordinary explanations for the anomaly, Anderson and Iris JPL collaborator Slava Turyshev chose the latter. A wise first step. You don't want to invent a new law of physics to explain a mere hardware malfunction.

Because the flow of heat energy in various directions can have unexpected effects, one of the things Anderson and Turyshev looked at was the spacecraft's material self—specifically the way heat would be absorbed, conducted, and radiated from one surface to another. Their inquiry managed to account for about a tenth of the anomaly. But neither investigator is a thermal engineer. A wise second step: find one. So in early 2006 Turyshev sought out Gary Kinsella, a JPL colleague who until that moment had never met either him or a Pioneer face to face, and convinced Kinsella to take the thermal issues to the next level. Last spring, all three men came to the Hayden Planetarium in New York City to tell a sellout crowd about their still-unfinished travails. Meanwhile, other researchers worldwide have been taking up the challenge too.

Consider what it's like to be a spacecraft living and working hundreds of millions of miles from the Sun. First of all, your sunny side warms up while the unheated hardware on your shady side can plunge to 455 degrees below zero Fahrenheit, the background temperature of outer space. Next, you're constructed of many different kinds of materials and have multiple appendages, all of which have different thermal properties and thus absorb, conduct, emit, and scatter heat differently, both within your various cavities and outside to space. In addition, your parts like to operate at very different temperatures: your cryogenic science instruments do fine in the frigidity of outer space, but your cameras favor room temperature, and your rocket thrusters, when fired, register 2,000 degrees F. Not only that, every piece of your hardware sits within ten feet of all your other pieces of hardware.

The task facing Kinsella and his team of engineers was to assess and quantify the directional thermal influence of every feature on board Pioneer 10. To do that, they created a computer model representing the spacecraft surrounded by a spherical envelope. Then they subdivided that surface into 2,600 zones, enabling them to track the flow of heat from every spot in the spacecraft to and through every spot in the surrounding sphere. To strengthen their case, they also hunted through all available project documents and data files, many of which hail from the days when computers relied on punch cards for data entry and stored data on nine-track tape. Without emergency funds from the Planetary Society, by the way, those irreplaceable archives would shortly have ended up in a dumpster.

For the simulated world of the team's computer model, the spacecraft was placed at a test distance from the Sun (25 AU) and at a specific angle to the Sun, and all the parts were presumed to be working as they were supposed to. Kinsella and his crew determined that, indeed, the uneven thermal emission from the spacecraft's exterior surfaces does create an anomaly—and that it is indeed a continuous, sunward change in velocity.

But how much of the Pioneer anomaly can be chalked up to this thermal anomaly? Some. Perhaps even most. But not all. The team's thermal model was based solely on the trajectory and hardware data from Pioneer 10, which displays a smaller anomaly than that of Pioneer 11. Not only that, the researchers have yet to calculate how the thermal anomaly varies with Pioneer 10's (let alone Pioneer 11's) distance from the Sun.

So what about the as-yet-unexplained "not all" portion? Do we sweep it under the cosmic rug in hopes that additional Kinsellan analysis will eventually resolve the entire anomaly? Or do we carefully reconsider the accuracy and inclusiveness of Newton's laws of gravity, as a few zealous physicists have been doing for a couple of decades?

Pre-Pioneers, Newtonian gravity had never been measured—and was therefore never confirmed—with great precision over great distances. In fact, Slava Turyshev, an expert in Einstein's general relativity, regards the Pioneers as (unintentionally) the largest-ever gravitational experiment to confirm whether Newtonian gravity is fully valid in the outer solar system. That experiment, he contends, shows it is not. As any physicist can demonstrate, beyond 15 AU the effects of Einsteinian gravity are negligible. So, at the moment, two forces seem to be at play in deep space: Newton's laws of gravity and the mysterious Pioneer anomaly. Until the anomaly is thoroughly accounted for by misbehaving hardware, and can therefore be eliminated from consideration, Newton's laws will remain unconfirmed. And there might be a rug somewhere in the cosmos with a new law of physics under it just waiting to be uncovered.

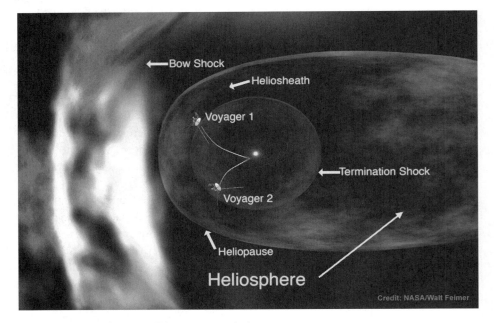

Image depicting distance of Voyager spacecrafts from our sun.

Go Boldly, Voyager[*]

By Tim Appenzeller
National Geographic, February 2006

The starship Enterprise it's not. Voyager 1 is a spidery contraption with less computer memory than some calculators and barely enough power to light three 100-watt bulbs. Yet this 1970s-vintage space probe will soon become humanity's first envoy to the stars.

Now hurtling at nearly 40,000 miles an hour well past the orbit of Pluto, Voyager has traveled farther than any other spacecraft. In late 2004 it passed a space boundary called the termination shock, a milestone near the outer limits of the solar system where the thin wind begins to collide with winds from interstellar space. In another ten years Voyager will space between the stars. "Interstellar space, for the first time!" exults Eric Christian, Voyager's NASA program scientist.

For as long as NASA funds it and its decades-old technology holds out, Voyager will continue its epic journey of exploration. Launched in 1977, Voyager 1 and its companion, Voyager 2, made a 12-year tour of the outer solar system, visiting planets and moons all the way out to Uranus and Neptune—places never reached before or since. Along the way it transformed our view of the solar system. "Voyager really was the transition from the basic nine planets to an incredibly mixed plethora of worlds," says Louis Friedman, executive director of the Planetary Society.

The probes have revealed that Jupiter has faint rings, that Saturn's rings—more intricate than anyone imagined—are peppered with moonlets, and that 900-mph winds churn Neptune's atmosphere. The moons of the outer planets had long been expected to be pitted and dead like our own, but Voyager found they were a study in diversity: seething with volcanic activity, swathed in hydrocarbon smog, crusted with ice floating atop what may be a hidden ocean. Says one scientist: "NASA in the 20th century is going to be remembered for two things: Apollo moon landings and Voyager."

Yet Voyager started out as a compromise. It was conceived in the early 1970s after NASA abandoned its plans to take advantage of a rare alignment of planets by sending a costly fleet of four spacecraft on a "grand tour" from Jupiter out to Pluto. The two Voyagers were designed to go no farther than Saturn.

"Of course, we all had our hopes that the mission could go on," says Ed Stone, the Voyager chief scientist since 1972. He and his colleagues knew that if they sent a spacecraft careering past Saturn at just the right angle, Saturn's gravity would fling it straight toward Uranus and Neptune. After Voyager 1 explored Saturn and its moons, perhaps Voyager 2 could be vectored toward more distant planets.

Stone and his colleagues got their wish; in 1981 NASA extended the mission. In the following years Voyager 2 dazzled with close-ups of Uranus, Neptune, and their moons. Leaving Saturn, Voyager 1 climbed out of the plane of the planets and headed directly toward interstellar space.

As the decades passed, both Voyagers stayed healthy in spite of brutal cold and a barrage of cosmic rays. The spacecraft had been built to survive intense radiation near Jupiter. "In a sense radiation is a very rapid aging effect," says Stone. After Jupiter, he notes, the additional aging over 28 years didn't make much of a difference.

As the planets dwindled behind it, Voyager 1 had a quiet journey until mid-2002, when it detected bursts of particles apparently sprayed from a nearby shockwave. On December 16, 2004, it reached the source. A sudden strengthening of the solar wind's magnetism indicated that the wind had slowed and piled up—just what was expected at the termination shock. Voyager 2, on a slower route out of the solar system, is expected to reach the shock in as little as two years.

In another decade Voyager 1 should finally cross the helio-pause, the last gasp of the solar wind, then sail out among the stars. With some 15 years left in its plutonium power source, it may still be alert and talkative. If cash-strapped NASA can keep finding 45 million dollars a year—a bargain compared with other missions—Voyager will give scientists on-the-spot reports from interstellar space.

It will be 40,000 years before Voyager 1 drifts past a neighboring star. For any aliens they might meet, both Voyagers bear a gift: a disc storing images and sounds from Earth. The technology, however, falls short of Star Trek standards. Each disc is designed to be played on a turntable, and a phonograph needle is thoughtfully included.

Cosmic Flock[*]

By Dan Cray
Time, March 20, 2008

Human beings have a habit of making traffic wherever they go. Give us a new means of transportation, and pretty soon highways, sea-lanes and airline routes are filled with vehicles. Now add to that deep space.

For all the attention that the shuttle, the space station and other manned spacecraft get, the real foot soldiers of space exploration have always been the unmanned ships—and right now they're enjoying something of a golden age. The U.S. currently has no fewer than 11 interplanetary probes scattered about the solar system; five are orbiting, roving or approaching Mars alone, and the others are targeting Mercury, the sun, Saturn and numerous comets or asteroids. One probe is heading for a never before rendezvous with Pluto, a destination it won't reach until 2015.

This spring three of the rugged ships stand out from the rest. Near Saturn, the Cassini orbiter, launched by the Jet Propulsion Laboratory, just executed a dramatic dive through an icy geyser that reaches 950 miles (1,530 km) into space from the Saturnian moon Enceladus, and there are plans to follow that up with even higher-risk maneuvers. In May NASA's Phoenix Lander will set down in Mars' arctic region in search of water ice. And later this month NASA and the European Space Agency will retire their Ulysses solar surveyor after a 17-year mission that has reframed our understanding of the sun.

All three missions have thrilled and surprised scientists—who pride themselves on knowing more or less what to expect. "I sit back with my mouth open, watching paradigms shift," says Linda Spilker, Cassini's deputy project scientist.

The orbiter's plume dive was responsible for some of that shifting. Passing just 120 miles (190 km) above the surface of Enceladus, Cassini sampled an icy exhaust that researchers didn't even know existed until the spacecraft spotted it three years ago. NASA expects to release detailed composition information soon, but the ice hints at subsurface water and the attendant possibility of life. Seven more close-brush flybys are in the offing, including one high-wire plunge that will drop the spacecraft a scant 15 miles (24 km) above Enceladus' surface. Says JPL's Spilker: "We're going to taste and sniff everything."

Before the orbiter attempts that maneuver, it will execute two flybys of the moon Titan, whose opaque orange atmosphere has been increasingly pierced by the spacecraft's radar. And this summer Cassini will make an unusually high orbit above Saturn's massive B ring, promising unique images of the ring, spread like an immense halo around the planet. The ship will also have the rare opportunity to observe the sun cross the plane of the ring from south to north, literally shedding light on the B ring's complex particle structure. "We want to know what a particle would look like if you could pick one up and hold it in your hand," Spilker says, "and we can do that by studying how they heat and cool."

Don't mention cooling to the researchers behind the Phoenix Mars Lander. Their ship will have just six months to sample and study the water ice at the Martian north pole before -200°F (-130°C) winter temperatures hit the region. "We last until the sun goes down. Then we freeze to death," says principal investigator Peter Smith, a planetary scientist at the University of Arizona, Tucson. Before it does, Phoenix Lander will probably offer a first look at actual Martian water ice rather than the dry water scars of millenniums past. To do that, the lander will use a digging arm and a suite of mineralogy instruments to hunt for salts, clays and other signs that liquid water is manipulating the soil. If Phoenix Lander hits its targets, this will be a big step toward later missions that will search for microscopic organic life. "Pay attention," Smith says, "because it's the polar region. No one's ever been there, and it's going to be fun."

Less glamorous but more sweeping than the half-year Phoenix mission was the long-running Ulysses mission, which took the first full measure of the sun's polar regions. If it swirls, floats or emanates near the sun, Ulysses studied it. The spacecraft discovered that the sun's magnetic field determines the regions that produce the solar wind, and ruffled more than a few scientists' feathers when it showed that a hot corona produces the fastest solar winds—exactly the opposite of prevailing theories.

Ulysses also tracked interstellar dust particles all the way from the sun to Earth, and in so doing helped map the planet's magnetic fields. The big surprise came when Ulysses stumbled on the tails of two comets and found that those feathery streams were more than 93 million miles (150 million km) in length. That's about the distance from the sun to Earth. "Totally unexpected," JPL project scientist Ed Smith says simply.

A diminishing power supply means the Ulysses mission ends on March 30, but the textbook rewrites will go on as fresh ships continue to take the place of old ones. The Lunar Reconnaissance Orbiter (LRO), which will launch later this year, will conduct the most comprehensive surveys of the moon the U.S. has ever attempted, using cameras that can spot an object as small as a football. The mission will help scout for landing sites, as NASA is holding fast to its plans to return astronauts to the moon by 2020. LRO will also hunt for signs of water ice on the moon, as well as help study the irregular lunar gravity field, caused by dense con-centrations of mass beneath the surface—the geological equivalent of lumps in oatmeal. Most dramatically, it will collect detailed images of all six Apollo landing sites, which have stood unseen for close to 40 years. "LRO's job is to open up the lunar frontier," says Jim Garvin, chief scientist at Goddard's Space Flight Center, where the craft is being assembled. "Right now we have a view from the 1970s, and here we are in the 21st century."

More missions to Mars are anticipated, including one that would return soil samples, possibly shedding fresh light on Martian life and allowing NASA to re-hearse the round-trip skills that would be necessary for a manned mission. And even as the new ships are readied, some of the great historic ones are still in flight. Voyagers 1 and 2, launched in 1977 on a grand tour of the outer planets, are now on their way out of the solar system, with the last breaths of solar wind at their backs. Remarkably, NASA may be able to stay in touch with them for up to 30 more years—meaning the granddaddy ships could remain online long after some of the newest ones have winked out. As traffic jams go, that's not bad.

6

Exoplanets:
The Search for Extrasolar Earths

Credit: NASA/Nigel Sharp

The Horsehead Nebula in the constellation of Orion.

Editor's Introduction

The focus of this book thus far has been the exploration of our own solar system. In the last chapter, the descriptions of the now-interstellar Pioneer and Voyager probes introduced the idea of exploring space *beyond* our solar system—something that scientists have been doing for decades through powerful Earth-bound telescopes, and more recently, with the orbiting Hubble Space Telescope. Using optical, radio, X-ray, and gamma-ray telescopes, astronomers have gathered a wealth of information about our galaxy, especially pertaining to stars that are in relative proximity to Earth. Until recently, however, they could only theorize about the presence of planets around such stars because no direct or indirect means of detecting them had been developed. Before the late 20th century, the existence of extrasolar planets, or exoplanets, was simply a centuries-old subject of scientific conjecture; no one could guess how common they were or how similar they might be to planets in our own solar system—or even if they existed at all.

The first exoplanets were found in the 1990s using an indirect method—by measuring the spectral "wobble" created when an extrasolar planet's gravity affects the motion of its parent star. Using this technological advancement in high-resolution spectroscopy, as well as other methods, astronomers have been able to confirm the existence of 307 exoplanets as of July 2008. The vast majority of these planets are massive gas giants similar in composition to Jupiter, the largest planet in our solar system. Many of these giants inhabit orbits unseen in this solar system—some very close to their suns, others in elliptical patterns—forcing many researchers to reevaluate their models of planetary development. Due to the present limitations of detection technology, no rocky planets the size of Earth have been spotted so far, though one, Gliese 581 d, the third planet circling the red dwarf star Gliese 581 in the constellation of Libra, seems to hold promise. It is slightly larger than Earth and near the so-called "habitable zone," the region of space around stars that is considered favorable to the development of life as found on our world. As detection methods are further refined, it is very likely that many more exoplanets of this size or smaller will be catalogued.

Scientists now estimate that 10 percent of stars comparable to our own sun have planets; how many of those planets reside in habitable zones is still unknown. For obvious reasons, habitable zones of distant planetary systems are now becoming the focus of many astronomers' studies. If life began here under specific conditions, in specific habitable zones, scientists argue, then it is likely it also began out there under similar conditions. The search for exoplanets has now become as much about the search for extraterrestrial intelligence as it is about scientific discovery. If exoplanets are discovered that seem to be mirrors of our Earth, then it is likely that scientists, particularly radio astronomers, will begin to concentrate on those regions of space, looking for any indication of intelligent life.

In this final section the collected articles will provide a general summary of the search for extrasolar planets. The first article, taken from the 2007 *Current Biography International Yearbook*, is on Michel Mayor, the Swiss astronomer who discovered the first exoplanet in 1995 with his student Didier Queloz. This biographical piece provides historical context as well as an overview of the methodology used in searching for exoplanets. In "Record Fifth Planet Discovered Around Distant Star," JR Minkel reports on the planetary system discovered around 55 Cancri, a sun-like star in the constellation of Cancer. In "The New Search for Distant Planets," Geoffrey W. Marcy introduces two other methods used to detect exoplanets—gravitational microlensing and transit—and describes "super-Earths," a new class of rocky planets that, in terms of mass, falls somewhere between Earth and Uranus. Margaret Turnbull details the parameters researchers use for establishing habitable zones around stars in "Where is Life Hiding?" In the next article, "Searching for Earth's History Among Earth-like Worlds," Lisa Kaltenegger describes how researchers look for signs of life on exoplanets. Finally, in "Where Are They? Why I Hope the Search for Extraterrestrial Life Finds Nothing," Nick Bostrom contends that if no intelligent life is found on extrasolar planets our civilization may breathe a collective sigh of relief—because that would mean that humanity has a greater chance for long-term survival.

Michel Mayor[*]

By Christopher Mari
Current Biography International Yearbook, 2007

Michel G.E. Mayor is primarily known for his research into extrasolar planets. In 1995 he and his student Didier Queloz discovered 51 Pegasi b, the first extrasolar planet to be found that orbits a still-living star. Since that time Mayor's teams have been responsible for finding approximately half of the 220 extrasolar planets that are now known to exist. Mayor and other astronomers have employed various indirect methods to detect planets that are light years from our own solar system and obscured from view by the brightness of the stars that they orbit. Most notably they have used spectrometers, which can detect a planet's wiggle as it passes in front of its star. While most of the planets found thus far have been gas giants, that is, Jupiter-like planets, technological advancements are now enabling scientists to detect smaller planets that may be similar to Earth and that could, theoretically, contain life. In 2007 Mayor's team discovered what most consider to be the first Earthlike planet found outside our solar system—a planet slightly larger than Earth, orbiting a red dwarf known as Gliese 581 in the constellation of Libra. While Mayor plans to continue his search for extrasolar planets, he has set his sights on a more fascinating discovery: life on another world. Mayor believes that scientists will have the ability to search for signs of life on other worlds within 15 to 20 years. He told Bradley S. Klapper for the *Washington Post* (April 25, 2007), "I feel comfortable with the idea of life existing elsewhere," although he has also noted that one cannot be certain that there is any to be found.

Michel G.E. Mayor was born in Lausanne, Switzerland, on January 12, 1942. While little has been published in English about his personal life, it is known that Mayor received his undergraduate degree in physics from the University of Lausanne and obtained his doctorate in astronomy at the Geneva Observatory, in 1971. He has had a long career as an astronomer, working at numerous observatories, including the Mullard Radio Astronomy Observatory at Lord's Bridge at the University of Cambridge, in 1971; the International Astronomical Union

(IAU), from 1988 to 1991; the European Southern Observatory (ESO) in Chile, from 1990 to 1992; the W. M. Keck Observatory at the University of Hawaii, from 1994 to 1995; and with the Swiss Society of Astrophysics and Astronomy from 1990 to 1993. During this time he worked on numerous papers, but outside astronomical circles, he is known only for his discovery of the first extrasolar planet, or exoplanet.

While people have speculated for centuries about the existence of planets outside our solar system, demonstrating that such planets exist became a possibility only in the late 20th century. Extrasolar planets cannot be seen from Earth, but the effect that they have on their stars can be, at least through a powerful telescope, because a planet does not orbit smoothly around its stars' centers. Instead, each body—planet and star alike—revolves around the center of gravity that they create in their planetary system, and the gravity of one body exerts a force on the other. Therefore, a star that has a planet, or planets, revolving around it does not move smoothly through its plane of existence; it wobbles as the force of gravity from its planet tugs at it. That wobble can be detected from Earth using a spectrometer, which measures a star's movement toward and away from the Earth. The first definitive detection of extrasolar orbital bodies came in 1992, when Aleksander Wolszczan and Dale Frail announced the discovery of objects orbiting the pulsar known as PSR 1257+12. (A pulsar is theorized to be a collapsed and, therefore, dead star. Its name is derived from the fact that it emits regular pulsating bursts of electromagnetic radiation, and the objects, which are called pulsar planets, that Wolszczan and Frail found are likely the leftover rocky cores of gas giants that had survived when their star went supernova, that is, when a star explodes.)

In 1994 Mayor and Queloz began using a 1.9-meter telescope at the Haute-Provence Observatory, in France, to find brown dwarfs—celestial bodies composed of gas that are larger than a planet like Jupiter but smaller than the smallest of stars. The duo planned to use a spectrometer to search 142 stars similar to our own Sun to detect the kind of wobbles in the spectrum that a brown dwarf would exhibit. (Brown dwarfs do not emit light, so they cannot be seen.) One of the stars on this list was 51 Pegasi, a middle-aged star in the constellation of Pegasus. While studying the data in February 1995, Queloz realized that 51 Pegasi was demonstrating the kind of spectrographic readings—a telltale wobble—that would suggest something was in orbit around it. However, the wiggles indicated that the object had a mass of approximately half that of Jupiter—far too small to be a brown dwarf. Both men assumed they had faulty instruments on their hands, not an extrasolar planet. "At the start we were extremely suspicious and looking for different explanations," Mayor told Jeffrey Winters for *Discover* (January 1996). "The first reaction was not to say, 'Oh! We have a planet!' At the start you say, 'Oh, something is wrong.' It's only after weeks or months that you start to be convinced."

A second look at the star, in March, proved that their instruments weren't malfunctioning. Mayor and Queloz, however, still wanted additional proof. Over the following four months, when 51 Pegasi was hidden behind the glare of our Sun, the two men determined the kinds of light signals they would see if a planet were orbiting the star. When the star reappeared, they saw those signals, and they announced their findings at an astronomy conference in Florence, Italy on October 6, 1995, stunning the scientific community. The planet, now named 51 Pegasi b, orbits a star that is very similar to our own Sun. It is nothing like the cold husks found orbiting pulsar PSR 1257+12, nor is it like any planet in our own solar system. Its mass is estimated to be about that of Jupiter's, and it orbits its sun in 4.2 days, heating up to between 1,200 and 1,800 degrees Fahrenheit during the journey. Mayor and Queloz could only speculate about the planet's composition, and they guessed that despite its close proximity to its star, 51 Pegasi b is likely a gas giant. Theories current at the time suggested that a gas giant could not form in close proximity to its parent star without being ripped apart by it. Upon hearing of the planet, Franco Pacini of the Arcetri Astrophysical Observatory, in Florence, Italy, asked incredulously, as Nigel Hawkes reported for the *London Times* (October 16, 1995): "How long could a planet last so close to the principal star without evaporating from the effect of the enormous quantity of energy it absorbed?"

Mayor had such questions himself and, before making his announcement, "asked several theoreticians how close Jupiter could be to the Sun before it became unstable. None of them knew. But the answer finally came from Adam Burrows of the University of Arizona. On hearing of the problem from an ex-student of Mayor's, he ran simulations of planets at all distances from a Sun-like star. Within 24 hours he came back with the news that anything farther away than 0.04 astronomical units would be stable. At 0.05 astronomical units from its star, Mayor's planet could comfortably exist," the *New Scientist* (June 15, 1996) reported. Nonetheless, many in the scientific community questioned Mayor and Queloz's findings—including David Gray of the University of Western Ontario, who asserted into 1997 that the planet didn't exist. Geoff Marcy of San Francisco State University and Paul Butler of the University of California at Berkeley independently confirmed the discovery 10 days after it was announced with the Lick Observatory's more sophisticated spectrometer and telescope. Marcy and Butler had also been searching for planets and, unknowingly, had evidence that Pegasi b—as well as several other planets—existed on their computers for some time. They had failed to recognize the bodies as planets because they had assumed that any extrasolar planetary system would abide by the theory of planetary formation that had been shaped by observations of our own solar system.

Theorists had assumed solar systems form when the cloud of gas, ice, and dust surrounding a protostar flattens into a spinning disk. In the outer part of the disk, ice remains solid, allowing planetary cores to form that are at least 10 to 20 times more massive than the Earth. These cores are so large that they have enough gravity to wrap themselves in deep gassy mantles, ultimately becoming gas giants similar to Jupiter or Saturn. Closer to a star, where it is warm enough to prevent large ice cores from taking shape, smaller rocky planets form through the agglomeration of dust particles. "Theorists had been building systems like ours, but the discovery of the pulsar planets and the 51 Pegasi planet shows that planets can form in ways that theorists haven't figured out," William Cochran of the University of Texas at Austin remarked, Robert Naeye reported for *Astronomy* (March 1996). Numerous other theories have been put forth to explain this unusual planet, the existence of which contradicted the planetary formation model. One theory postulated that the planet, which Mayor dubbed a "Hot Jupiter" owing to its size and proximity to its sun, had formed farther away from 51 Pegasi but had migrated inward.

Since the discovery of 51 Pegasi b, Mayor and his research team have found other extrasolar planets, as have many other astronomers. Those planets found in the months after Mayor and Queloz made their first announcement were very similar to 51 Pegasi b: their masses ranged from .44 to 10 times that of Jupiter's, and 50 percent of them orbited their parent stars quite closely. Some of these planets have also had very eccentric elliptical orbits, a revelation that also illustrates the limits of the standard planetary formation theory, which supposed that planets should have mostly circular orbits. In 1998 Mayor and Queloz announced that they had discovered a planet orbiting Gliese 614. It had a mass at least 3.3 times more than Jupiter's and circled its sun every 4.4 years at a distance of 2.5 astronomical units—much closer to the distance one would expect a gas giant to be orbiting its sun. However, such an exoplanet proved the exception, not the rule. By 2000 some 28 extrasolar planets had been discovered and verified; by April 2007, that number had increased to 220. Owing to the number and variety of those planets that had been discovered, a more inclusive planetary formation theory began to emerge in the early part of this century. New computer models depicted situations in which protoplanets, including gas giants, might move towards a parent star instead of away from it. With the right conditions a planet might even remain in a stable orbit close to the star, though more often than not, such large planets would be pulled apart by the parent star's gravimetric pull.

In late 2002 Mayor and Queloz announced that they had discovered another 12 extrasolar planets, some of which belonged to the same solar system. Early the following year, Mayor's High Accuracy Radial Velocity Planet Searcher (HARPS) came on-line at the ESO, in Chile. This device was developed to help Mayor find planets more like our own, something that potentially placed scientists one step closer to finding extraterrestial life. The 220 planets that had been found by early 2007 had what many called the "Goldilocks problem"—that is, they are all too hot, too cold, or too massive and gaseous to support life as we know it. On April 25, 2007, Mayor announced that his team had found a smallish, rocky planet in a zone that was close enough to its own star that liquid water could exist on its surface. By using the ESO's telescope, along with HARPS, Mayor's team uncovered a planet orbiting Gliese 581, a red dwarf that is in the constellation of Libra and is about 20.5 light years (or 120 trillion miles) from our solar system. The planet is 14 times closer to its sun than our Earth is to the Sun, and its gravity is likely much stronger than that found on Earth. Though the planet, dubbed Gliese 581 c, is very close to its star, which it orbits in 13 days, its temperatures are not extremely hot because a red dwarf is considerably cooler and dimmer than our own Sun. Scientists theorize that the planet's average temperature could be anywhere from 32 to 104 degrees Fahrenheit and thus could have water in liquid form on its surface, though verification is impossible at present. "We do not have any reason to believe that life exists on that planet," Mayor told Nell Boyce for the Morning Edition on National Public Radio (May 8, 2007). "We can only say that we have the temperature to permit the development of life. I would say it's one very inter-esting step in a long process going in the direction to having some major discovery related to life in the universe."

In recognition of his achievements Mayor has received numerous accolades. Noteworthy among his early prizes was the 1983 Prize of the Académie française des Sciences. Following his discovery of 51 Pegasi b, he was awarded, in 1997, the medal of the Commission of Bioastronomy by the International Astronomical Union (IAU); the Swiss Confederation's 1998 Prize Marcel-Benoist; the Astro-nomical Society of France's 1998 Janssen Prize; and the Observatory of the Côte d'Azur's 1998 ADION Medal. He also received the Balzan Prize, in 2000, the Albert Einstein Medal, in 2004, and the Shaw Prize in Astronomy, in 2005.

Mayor is currently associated with the Department of Astronomy at the Uni-versity of Geneva and has been the director of the Geneva Observatory since 1998.

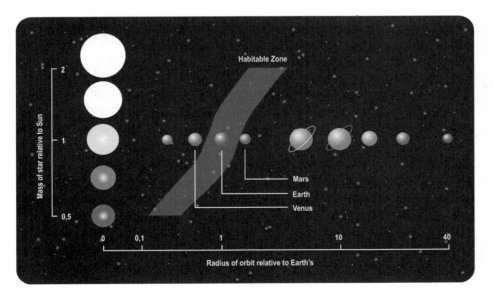

Habitable zone according to the size of the star.

Record Fifth Planet Discovered Around Distant Star[*]

By JR Minkel
Scientific American, November 6, 2007

Astronomers have spotted a record-setting fifth planet orbiting the sunlike star 55 Cancri, 41 light-years away in the constellation Cancer. Researchers say the planet, a "mini-Saturn" of about 46 Earth masses, lies fourth out from the star in a large gap between the third and fifth planets, placing it squarely in the estimated habitable zone around the star where water might remain liquid, according to the group's report, accepted for publication in *The Astrophysical Journal*.

Although the planet's size implies that it is a ball of hydrogen and helium gas incapable of supporting pools of liquid water, the finding raises the possibility that additional, earthlike planets might be discovered around it.

"This discovery of the first ever quintuple planetary system has me jumping out of my socks," says group member and veteran planet hunter Geoffrey Marcy, an astronomer at the University of California at Berkeley. "We now know that our sun and its family of planets is not unusual."

One of the first stars discovered to harbor an extrasolar planet or exoplanet, the 55 Cancri system has come to resemble a jumbo version of our own solar system. Its five planets all seem to orbit along relatively circular paths, and the farthest planet out, a gaseous behemoth the size of four Jupiters, revolves at roughly the same distance that separates Jupiter from the sun.

55 Cancri's innermost planet, weighing in at more than 10 earth masses—meaning it could have a rocky or icy core—lies closer to its star than Mercury does to our own. The new planet sits at 0.8 earth-sun distances (astronomical units) from the star, or roughly the distance between Venus and the sun. Before this discovery, researchers knew of only one other four-planet system, Mu Arae, and several three-planet systems.

Astronomers have uncovered 55 Cancri's planets one by one during 18 years of repetitive measurements at the Lick and Keck observatories in California and Hawaii, respectively. Researchers were looking at the star's Doppler shift, the change in the wavelength, or color, of its light as it moved toward and away from Earth. A star tugged by an orbiting planet will wobble slightly, which can be detected as a regular shift in the star's color corresponding to the time the planet requires to complete an orbit.

Multiple planets imprint multiple overlapping shifts, which require time—and mathematical modeling of possible planetary arrangements—to tease apart. 55 Cancri's outer planet, for example, has an orbital period of 14 years, and was therefore only discovered in 2004. The latest planet was even trickier to identify. "For me personally," says astronomer Debra Fischer of San Francisco State University, the study's lead author, "this was one of the more annoying stars. It resisted mathematical modeling because of this extra planet we finally have extracted."

The study authors are scanning several thousand other stars for exoplanets, but most of them haven't been scrutinized for as long as 55 Cancri, suggesting that more systems with five or more planets are lurking in plain sight of telescopes, says David Charbonneau, professor of astronomy at Harvard University, who was not involved in the study. "The excitement is, yes, there may be gaggles of planets around other stars in their survey as well."

Fischer says she expects 55 Cancri to harbor additional, smaller planets in the large remaining gaps around the new find, given that our own solar system is so densely packed with planets. However, its large outermost planet could have long ago swept that vicinity clean of planetary material, notes planetary scientist Jonathan Lunine of the University of Arizona at Tucson.

Future ground- and space-based experiments should have the power to discover Earth-size planets, which may be anywhere, Lunine says, but current technology is still too limited to spot them.

The New Search for Distant Planets[*]

By Geoffrey W. Marcy
Astronomy, October 2006

For centuries, friends have gathered for lunch and conversation at outdoor cafés in northern Greece. Under the hot Sun, surrounded by appetizers and rounds of ouzo, conversations have often meandered toward the heavens, sparking heated debates about the existence of other fertile worlds. If those worlds also harbor citizens, might they also gaze toward the stars? Such café philosophers flourished in 400 B.C., when Democritus taught his students about the possibility of other habitable worlds of great diversity, and of the likelihood of life on them. Modern astrophysics is finally poised to answer the Greek philosophers' questions about other worlds and the possibility of life on those worlds. In the past 10 years, some 200 planets have been discovered around nearby Sun-like stars.

NEW WORLDS FOUND

Michel Mayor and Didier Queloz of the Observatory of Geneva in Switzerland discovered the first exoplanet orbiting a main sequence star in 1995—a Jupiter-size object orbiting Sun-like star 51 Pegasi. Their findings appeared in the November 23, 1995, issue of *Nature*. Within the next 3 months, Paul Butler and I discovered the second and third exoplanets orbiting Sun-like stars 47 Ursae Majoris and 70 Virginis, respectively, while at San Francisco State University. We reported our discovery in the June 1996 issue of the *Astrophysical Journal*. These discoveries gave birth to the new field of extrasolar planetary science. The parade of discoveries has continued nonstop since then.

Astronomers have discovered most of the known exoplanets by detecting the reflex motion of their host stars as they were yanked gravitationally by their orbiting planets. As the star wobbles toward us and away, its emitted light waves are alternately compressed and stretched. Earthbound telescopes equipped with

* Reproduced by permission. © 2006, *Astronomy* magazine, Kalmbach Publishing Co.

spectrometers that spread the starlight into its composite colors can detect this Doppler effect. However, only planets the size of Jupiter, Saturn, or Neptune have been detected using the Doppler method because only such giant planets have enough gravitational pull to yank the stars around by a detectable amount.

The giant planets found around other stars have startling properties. About 90 percent of them travel in elongated, eccentric orbits more akin to the comets in our solar system. Therein lies a possible origin of those eccentricities. Other planets or the exoplanet-forming protoplanetary disk can exert a gravitational force that pulls an exoplanet out of its original circular orbit. Similarly, Jupiter's strong gravitational field sling-shot comets in the Oort Cloud into distant, elongated orbits.

Another oddity is that many extrasolar giant planets—including the first ones found—orbit closer to their stars than the Earth is to the Sun (1 AU). These tight orbits defy conventional planet-formation theories in which giant planets form in the cooler, slower-moving outer reaches of a planetary system, where gas can settle gently onto rocky cores. Exactly how giant planets migrate inward and then park themselves close to their host stars remains a puzzle.

THE SMALLEST EXOPLANET

Improvements in Doppler technology allow us to measure stars' velocities to within 2 miles per hour (3km/h), or human walking speed, which has permitted discovery of smaller and smaller planets. Our team detected the smallest planet yet found around a nearby star, Gliese 876: a 10th-magnitude red dwarf that is only 15 light-years away. This fascinating star has two Jupiter-size planets with orbital periods of 30 and 61 days in a resonance that causes them to gravitationally shepherd each other, keeping them locked in that 2:1 period ratio.

As we monitored those two Jupiter-size exoplanets with the world's largest telescope, the Keck I in Hawaii, we were surprised the star exhibited an additional wobble not explainable by the two exoplanets. We found that a third planet of 7.5 Earth-masses and a remarkably short orbital period of only 1.9 days causes the additional wobble. Normally, the Doppler method reveals only a planet's minimum possible mass, which is limited by the unknown tilt of the orbital plane. But, in this fortuitous find, the two Jupiter-size exoplanets exert gravitational forces on each other that we detect, thereby establishing their orbital plane that, in turn, sets all the planet masses precisely.

A Doppler team led by Christophe Lovis in Geneva recently discovered a system of three Neptune-size planets with minimum masses of 10, 12, and 18 Earths orbiting HD 69830, a 5.95-magnitude, spectral-type KO star about 15-percent less massive than the Sun. The star is just visible to the naked eye.

Astronomers using the Spitzer Space Telescope discovered an asteroid belt outside the system of the three Neptune-size planets. They examined infrared-light emission from the dust to make the discovery. Thus, this system has properties similar to our solar system, which indicates planetary systems like ours may be common.

TWO MORE DETECTION METHODS

Adding to the planetary parade are two marvelous new detection techniques: gravitational microlensing and transit. Gravitational microlensing occurs when light from a distant background star bends around a planet and briefly amplifies ("lenses") the star's light. Two planets of 5 to 15 Earth-masses were discovered recently by this method with orbital distances between those of Mars and Jupiter.

About 8 planets have been discovered using the transit method, during which a planet crosses in front of and dims its host star. This technique provides a direct measure of an orbiting planet's diameter. The bigger the planet, the more starlight it blocks. Combined with the Doppler technique, which gives the masses of the planets, we can determine the planets' densities by dividing the planet's mass by its volume. Giant planets have densities near that of water, 1 gram per cubic centimeter (about the volume of a sugar cube), which is similar to Jupiter and Saturn's. Thus, most exoplanets found so far are gaseous objects.

SEARCHING FOR SUPER-EARTHS

The exoplanets of lowest mass point toward a new class, called super-Earths—exoplanets with masses greater than 1 but less than 14 Earth-masses. Interestingly, our solar system is devoid of planets in this range. Super-Earths represent the next great terra incognita to be explored—worlds larger than Earth but smaller than Uranus.

If super-Earths actually form, would they be rocky like the terrestrial planets, gaseous like Jupiter and Saturn, or icy and gaseous like Neptune and Uranus with a large rocky core? No one knows.

But observations give early clues about the existence and composition of super-Earths. Nature appears to make many more Saturn-mass planets than Jupiter-mass ones, and Neptune-mass planets seem even more abundant. Extrapolation suggests there might be yet more of the lower-mass super-Earths than all the giant planets combined.

Some known gaseous exoplanets may have rocky super-Earth cores. The best example of a likely rocky core is inside the transiting exoplanet HD 149026b, discovered by Debra Fischer of San Francisco State University. This planet has a 15-percent larger mass than Saturn but a 10-percent smaller radius. How could a planet be more massive, yet smaller in size? The apparent compaction of this Saturn-like planet can be explained only by its having a high concentration of heavy atomic elements that increases its gravity and, hence, final density.

Saturn has a rocky core 20 times Earth's mass. The Saturn-like exoplanet HD 149026b must have an even greater concentration of rocky and iron material than Saturn, and calculations by Peter Bodenheimer and collaborators suggest the rocky core is 50–70 Earth-masses. That whopper of a super-Earth is buried under an envelope of 30 Earth-masses of hydrogen and helium gas and strongly suggests nature has no trouble making such enormous cores of silicates and iron. Indeed, exoplanets predominantly orbit stars rich in heavy elements, indicating rock and iron play strong roles in planet formation.

HOW WOULD A SUPER-EARTH FORM?

Theorists have their own predictions about super-Earth formation. They predict rocky planets under 15 Earth-masses form when dust particles in the protoplanetary disk clump together. As dust sticks together and grows, like dust bunnies behind your bed, the largest blobs grow faster and faster as their large sizes allow them to collide frequently with ever more dust. These dust-and-ice blobs act like a thick sea of "planetesimals." The largest planetesimals begin gravitationally attracting smaller planetesimals, causing large ones to grow even faster.

Planet-formation models by Peter Goldreich, Yoram Lithwick, Re'rm Sari, Scott Kenyon, and Benjamin Bromley all agree about the next planet-formation steps. (The scientists are from Caltech, UC Berkeley, Caltech, the Smithsonian Astrophysical Observatory, and the University of Utah, respectively.) Rich planetesimals get richer and, after about 10 million years, Mars-size planets form. Dozens of such planets may orbit a typical young star.

Mars-size planets perturb each other gravitationally, causing orbits to cross and, eventually, collide violently with each other. Crushing upon impact, they stick and grow into Venus-or Earth-size planets after some 20–40 million years. Glancing impacts by two such planetesimals can result in ejected magma, which can form a large moon, as presumably created our Moon.

Theory predicts protoplanetary disks that are particularly rich in silicate dust or ice particles will give birth to especially large rocky planets of 5–10 Earth-masses. After all, why should Earth represent the largest rocky planet possible in the universe? Super-Earths must have massive iron-nickel cores compressed to somewhat higher densities, and they would have massive mantles heated by radioactive uranium, warming the interiors to tens of thousands of degrees Fahrenheit.

A FINE LINE BETWEEN ICE AND WATER

Many super-Earths will form within the disk's "ice line," located about 2 AU from the star, near our asteroid belt. The temperature within that distance is too high for ice crystals to exist. Super-Earths formed within the ice line will be composed mostly of silicates, iron, and nickel, as is Earth.

However, water could be delivered to such inner rocky planets by ice-rich asteroids, which astronomers think are abundant in our own solar system. A Jupiter-like planet would gravitationally perturb the hydrated asteroids into Earth-crossing orbits. The asteroids would eventually slam into the terrestrial planets and bring water to them. In this way, rocky planets can acquire various amounts of water, depending on how perturbative the Jupiter-like planet happens to be (due to its mass or orbital eccentricity).

Thus, some rocky planets might be nearly devoid of water—mere desert worlds without lakes, rain, or oceans. Other planets could receive 10 times, or even hundreds of times, the number of hydrated asteroids that Earth did, covering those worlds completely with a thick ocean. After all, if Earth had just twice its volume of ocean water, hardly any land would poke above sea level.

However, some planets may form beyond the ice line, in the cold outer reaches of the protoplanetary disk. There, ice particles dominate, comprising 3 times more mass than silicate dust particles. Protoplanets formed there will be composed of as much water as rock. The moons around Jupiter and Saturn are just such water worlds, albeit frozen and small. Callisto, Dione, and Enceladus are such worlds, having as much water-ice as rock. If any of those moons had migrated closer to the Sun, the added heating from sunlight would have melted them into liquid water worlds. Many Earth-mass-and-above planets located within 1 AU of their host stars may be such water worlds, with thick oceans and no continents.

PLANET QUEST IN FULL SWING

The coming decade will see the Greek philosophers' planetary quest met with various instruments designed to detect super-Earths around Sun-like stars. In 2008, the Kepler mission is scheduled for launch. It will be a 3-foot (1 meter) telescope with a 95-megapixel camera consisting of 42 charge-coupled devices (CCDs) to monitor 100,000 stars of roughly 12th magnitude in a field of view of 105 square degrees.

The stars' brightnesses will be monitored with a precision of one part in 100,000 to detect dimming caused by transiting Earth-size planets. The Kepler mission will give us a statistical measure of the occurrence of planets from Earth-size to super-Earths to Neptune-size orbiting within 2 AU of normal stars. For the first time, we may know how common other rocky worlds are.

NASA's Space Interferometry Mission (SIM) PlanetQuest is designed to detect planets of 3 Earth-masses or larger that orbit stars within 20 light-years of Earth. SIM PlanetQuest will use two telescopes separated by 295 feet (90m) as an interferometer combining visible light waves at a common focus. The enhancement or cancellation of light-wave crests and troughs from each telescope determines a star's position in the sky with an incredible accuracy of 1 millionth of an arcsecond. Thus, SIM PlanetQuest will hunt for the wobbles of stars pulled around by super-Earths orbiting between 0.1 and 2 AU from their host stars—where liquid water should be. Unfortunately, NASAs recent budget mandate, which emphasizes putting people back on the Moon, and fixing the space station, has delayed SIM PlanetQuest's launch date until at least 2015.

Meanwhile, the tried-and-true Doppler technique, responsible for locating some 185 exoplanets already, will get a boost with the completion of the "Rocky Planet Finder" Telescope at Lick Observatory. This 7.9-foot (2.4m) Hubble-size telescope will hunt for rocky planets 365 nights per year. The Rocky Planet Finder Telescope will robotically measure stars' wobbles with a precision of 2 miles per hour (3 km/h), using a specialized, high-resolution spectrometer designed by Steve Vogt of the University of California, Santa Cruz, famous for his construction of two previous planet-hunting spectrometers. Nightly measurements will detect rocky, 1–15 Earth-mass worlds orbiting within 0.5 AU of their stars by the gravitational wobbles they impart to their host stars.

The Rocky Planet Finder Telescope is scheduled for completion late next year. But funds for its operation are still needed, and the telescope has no official name. Nonetheless, the detection of rocky planets around nearby stars portends a new era in planet-hunting.

By targeting 200 nearby stars with the Rocky Planet Finder Telescope, planets with rocky surfaces and lukewarm temperatures will be detectable. Such warm planets would likely have surfaces with liquid water, serving as the solvent for biochemical reactions necessary for life to arise. Detecting habitable planets within 20 light-years of Earth would represent a milestone in science and answer a question posed by the ancient Greeks. Finding habitable planets would answer one of the most central questions of humanity: Do other habitable worlds exist? If so, what is their occurrence rate for stars slightly less massive than the Sun?

SEARCHING FOR INTELLIGENT LIFE

One great value of nearby habitable worlds is the ability to check them for signs of intelligent life. We will use Earth's major radio and optical telescopes to search for regular, pulsing signals from our newly discovered habitable worlds; only a technological civilization could produce regular signals. We plan to use the Allen Telescope Array (ATA), the new radio telescope by the University of California, Berkeley, and the SETI Institute, for this purpose. The ATA, its construction already under way, will enable high-sensitivity monitoring of our newly discovered habitable worlds to search for radio signals from intelligent civilizations. We will also search for extraterrestrial-intelligence signals in optical and near-IR wavelengths.

Perhaps within our lifetimes, or those of our grandchildren, we may sit at outdoor cafés along the Mediterranean neither wondering if other habitable worlds exist nor pondering if anyone lives on them. Instead, we may already know that we are not alone. If so, the lunchtime debates may focus on what constitutes "civilized" behavior in the eyes of our galactic neighbors.

Where Is Life Hiding?[*]

By Margaret Turnbull
Astronomy, October 2006

The search for life on other planets revolves around the concept of habitability. What makes our own Sun such an excellent parent for life? Our planet has remained habitable for life for some 3.7 billion years. So, what is it that allows a planet—and its parent star—to support life?

We know life will take hold if given a chance. This is a lesson biologist have learned about Earth, where organisms prosper in ice, rock, steaming geysers, and acidic pools. We also know life changes its environment, often creating new niches for life where there were none before. Ultimately, it seems all of life's basic requirements come down to one thing: liquid water. Life needs other things, too, like construction materials (such as carbon, nitrogen, and oxygen) and an energy source (starlight, a planet's internal heat, or chemical energy). But I submit that any environment with pools of liquid water will likely have these things, too.

Part of the search for living worlds beyond the solar system involves the idea of a habitable zone around each star. This is a region where the temperature is right for the presence of liquid water on an earthlike planet. This circumstellar habitable zone is only one of the potential abodes for thriving ecosystems. But at our early stage of understanding other planetary systems, the circumstellar habitable zone is the only nook we can study from afar.

STABILITY: THE FIRST OBSTACLE

While we still have much to learn about the Sun's habitable zone, we do know Earth lies within it and has been there for billions of years. Venus—our sister planet in terms of size and mass—is 30-percent closer to the Sun than Earth and doesn't lie within the habitable zone. Venus is too hot for life and likely lost its water early in its history as oceans were boiled off by the searing sunlight.

Mars, 50-percent farther from the Sun than Earth, is colder but, nevertheless, had liquid water flowing over its surface in the relatively recent past. Mars is probably within the Sun's habitable zone, and if the planet were just a tad more massive, its gravity could hold a substantial atmosphere.

A habitable zone that lasts billions of years is an important factor in the search for planets with life. As of now, we know nothing about the presence or absence of terrestrial planets in the habitable zones of the Sun's neighbors. However, we know a lot about the stars themselves, and we can see which would make good parents.

A good parent star is one with a habitable zone that remains in the same place for billions of years. The second constraint is nothing should prevent the formation of terrestrial planets. Not all stars pass these hurdles.

The first requirement—habitability over billions of years—puts strict constraint on several stellar parameters that are easily observed. Young stars are not the best places to look. Not only has life had less time to develop, but, for the first billion years or so, asteroids and comets bombard the system, frustrating life's efforts to survive. It turns out that stars—like adolescents entering adulthood—go through a significant decrease in flaring and other chromospheric activity after an age of 3 billion years. The Sun is one such example of a star that significantly decreased its flaring activity at an age of 3 billion years. Whether this newfound calm helps life form is unclear, but, at the very least, this lets us identify and rule out the youngest stars from our searches.

Long-term habitability also points us to a limited range of stellar masses. A Sun-like star has a stable hydrogen-burning lifetime of about 10 billion years—plenty of time for civilizations to evolve. Solar-type stars are also bright enough that the width of their habitable zones can accommodate two or more planets' orbits.

BEWARE EXTREMES

As many as 90 percent of the stars in our galaxy are small, dim objects called type-M dwarf stars—with masses at most half of the Sun's. Given their large numbers, we would like these M stars to host habitable planets. Because of their low masses, M stars have incredibly long lives—and, thus, long periods of near-constant luminosity. Around M dwarfs, life would have hundreds of billions, even trillions, of years to develop.

However, M stars have long been suspect in terms of their suitability for biology. Not only are their habitable zones narrow (about 1/10 the width of the Sun's), but these habitable zones are so close to the stars (1/4 Mercury's orbit, or 1/10 Earth's orbit) that any planets orbiting there would certainly be tidally locked. This means one side of a planet always faces its star. So, instead of getting an even roasting, one side of the planet would be a blazing noonday Sahara, while the other is a freezing midnight Siberia.

Initially, scientists thought such a situation would be unstable, and moisture in the planet's atmosphere would condense into a giant icecap on the night side. This in itself would be bad for life, but what is potentially worse is M stars are famous for flares. All stars flare occasionally, including the Sun. But M-star flares can temporarily increase the star's luminosity by a factor of 100 or more, blasting out high-energy radiation and fast-moving particles that would dismember any nearby DNA molecules. Combine these outbursts with the habitable zone's close distance, and we're talking about putting a planet in an ultraviolet sterilizer.

Even with M dwarfs' close orbits and solar flares, all is not lost for these stars. If a tidally locked planet is massive enough to hold a substantial greenhouse atmosphere, one not much thicker than Earth's, atmospheric circulation could keep conditions fairly mild all around the planet. Add a little oxygen, which builds up when water molecules dissociate, and all the worst flares could do is drive ozone creation in the planet's stratosphere. Ozone, in turn, shields the surface from the harsh radiation of solar flares. Moreover, not all M-class stars flare radically. While M stars are perhaps not the best real estate in the galaxy, most scientists believe planets around M stars can possibly support life. The SETI (Search for Extraterrestrial Intelligence) Institute's target list contains many M stars that change brightness by less than 3 percent.

When it comes to stars that are just a few times the Sun's mass, however, the picture is bleak. These stars burn so brightly they use all of their hydrogen fuel before the planets around them even finish forming. These stars swell into giants and fuse helium, carbon, nitrogen, oxygen, and other elements, synthesizing heavier elements all the way up to iron. That's as heavy as elements in normal stars can go. Then these stars die in explosions that create the rest of the natural elements and seed the galaxy with them.

Impressive? Yes. Good for life? No—or rather, not good for life today. We must never forget the first batches of massive stars created iron, nickel, carbon, oxygen, and the other building blocks of planets and life. These massive stars play a crucial role in the circle of life.

HOW MANY STARS DO YOU ORBIT?

After stars pass the mass, age, and flare-activity tests, we're sure they can provide roughly static habitable zones for long periods of time. But we still have our second criterion to consider: that nothing prevents planets from orbiting in those static habitable zones.

This criterion can be broken down into two components, the first of which is what astronomers call multiplicity. The Sun is a single star, which is slightly unusual: About two-thirds of otherwise Sun-like stars come in multiples. I don't mean to suggest that two, or more, stellar parents is necessarily a bad thing, but understanding the stars' orbits is crucial. For example, a second star could:

- orbit so close to its primary that the habitable zone encircles both stars—any earthlike planets will simply enjoy two sunsets per day;

- sweep through or come very close to the habitable zone, gravitationally perturbing the orbits of any planets there and possibly ejecting the planets altogether;
- orbit far enough away that it does not perturb the habitable zone but introduces a randomly changing luminosity that would interfere with the climate of any planets;
- have such a large orbit that any planets in the habitable zone are completely unaffected.

In the fourth case, both stars have "safe" habitable zones, which makes these systems doubly intriguing. Only small ranges of separation between the stars cause problems.

Most binaries appear safe, falling either into the first or fourth scenario. Triple and quadruple star systems sound more exotic, but they are just as likely as binaries to be safe harbors for life. Multiple systems are hierarchically configured: A triple almost always consists of a close double orbited by a distant third; a quadruple consists of two pairs; a quintuple of two pairs plus a distant third around one of them; and so on.

A STAR'S STUFF IS IMPORTANT

Thanks to the diligent efforts of planet hunters, we know about some 200 planets orbiting more than 160 of the Sun's stellar neighbors. Each of these planetary systems has to be assessed for stability, and a fair number can be ruled out because the giant planet interferes with the habitable zone. Most of the giant planets are on elliptical orbits and have gravitational influences that extend beyond their own orbits. Where such a gravitational influence extends into a star's habitable zone, it is unlikely any earthlike planets could maintain stable orbits. There are several interesting cases, however, (like Mu [μ] Arae) where giant planets orbit within the habitable zone. If these giants have big moons, similar to Jupiter's and Saturn's satellites, the moons could be adequate hosts for life.

The star's composition is the second factor that could prevent planets from orbiting in the habitable zone. We know stars and their planets form at the same time out of the same material. As density waves sweep through the galaxy, they compress gigantic clouds of gas and dust. Cloud fragments then collapse gravitationally into stars and planetary systems. But not all clouds are the same; some are deficient or enriched in elements heavier than helium (astronomers call these elements metals) depending on the local history of massive-star explosions. Thus, some stars are metal-poor, and some stars are metal-rich compared to the Sun.

This is important because astronomers have found metal-rich stars are more likely to have close-orbiting giant planets. Astronomers don't yet know how a star's metallicity correlates to the presence or absence of terrestrial planets, but it makes sense that iron-bearing planets like ours wouldn't form out of clouds that had no iron to begin with.

On the other hand, high-metallicity stars have close-orbiting giant planets because after the stars formed, there was still lots of material orbiting the new stars. That material aggregated into giant planets.

LOCATION, LOCATION, LOCATION

Finally, we need to question the habitability of the Milky Way Galaxy itself. Are all of the Milky Way's neighborhoods equally good for life? Not likely. Our Sun inhabits the galactic backwoods, where stars are separated, on average, by a few light-years and rarely pass within a comet's throw of one another. Stars in more crowded areas, such as the central galactic bulge, globular clusters, and spiral arms, are more likely to have close approaches from stellar neighbors. Such visits could destabilize planetary systems in the process.

The Sun probably formed in a loosely associated cluster of newborn stars. It may be that a few of our minor planets beyond Pluto (like Sedna) originally belonged to another star. But the Sun is a rural star now, and it will likely remain so for a long time. Although scientists don't fully understand the Sun's orbit around the galactic center, it seems the Sun orbits at a special location in the galactic disk called the co-rotation zone. This is where the galaxy's spiral pattern and the stars between arms move at roughly the same speed.

Closer to the galactic center, solitary stars move faster than the spiral pattern and, therefore, continually pass through the spiral arms. Spiral-arm regions of massive-star formation and high-energy radiation could fry life-forms. And regions of interstellar gas and dust could interfere with a planet's climate, by seeding the formation of rain clouds. By sticking to the middle lane, the Sun can remain suspended between spiral arms for billions of years.

Biology is a cosmic balancing act, with the best homes nestled in special nooks around special stars. These stars orbit special parts of the galaxy. Does it go beyond that to special pockets of galactic clusters throughout the universe?

If there is one thing life on Earth suggests, it's that there is an exception to every rule. Scientists have found life in the most unexpected places on our planet, which is promising for those searching for life outside of our solar system. And now that we have a better idea of where to look—from a star's composition to its age to its orbit—we are that much closer to finding out if were not alone.

POSSIBLE NOOKS FOR LIFE

*The following places might harbor worlds that could possibly have water—
the most important ingredient for life as we know it.*

- **Terrestrial planets in a star's habitable zone**

  The "Goldilocks" zone, not too hot or too cold, around a star could have earthlike planets with temperatures that allow liquid water on the planets' surfaces.

- **Terrestrial moons in a star's habitable zone**

  Giant planets orbiting within a star's habitable zone could have big moons (like Saturn's Titan) with large amounts of liquid water.

- **Tidal habitable zones around giant planets**

  Giant planets in the cold outer regions of a planetary system could have moons that are continually tidally stretched (like Jupiter's Europa) as they orbit their massive parents. This generates heat, which could provide for liquid water.

- **Free-floating terrestrial planets warmed by internal radiation**

  As planets form, some are undoubtedly ejected into interstellar space. The galaxy could be full of these free-floating planets whose interiors, heated by decay of radioactive elements, stay warm enough for liquid water to exist for billions of years.

- **Floating oceans within giant planets**

  Some giant planets could have layers in their interiors where the temperature and pressure are just right for liquid water.

TOP TEN PLACES TO LOOK FOR LIFE

Jill Tarter, director of the SETI (Search for Extraterrestrial Intelligence) Institute's research center, asked me to list the highest-priority targets for the upcoming Allen Telescope Array (ATA). At the time, I questioned astronomy's connection to biology. But as I became further involved in this project, the connection became ever more apparent.

My target list for SETI eventually evolved into a catalog of stars that may harbor habitable planets like our own. These stars are our best bets for finding earthlike planets using both radio telescopes (like the ATA) and NASA's Terrestrial Planet Finder (TPF) mission (now facing budget cuts).

STELLAR TARGETS FOR SETI SIGNAL SEARCHES:

- **1 Beta Canum Venaticorum: Yellow G-type star, slightly larger than the Sun; 26 light-years away, has a possible distant companion.**
- **2 HD 10307: Yellow G-type star; slightly hotter than the Sun; 41 light-years away; has a companion.**
- **3 HD 211415: Yellow G-type star; slightly cooler than the Sun; 44 light-years away; in a binary system.**
- **4 18 Scorpii: Yellow G-type star; slightly hotter and brighter than the Sun; 46 light-years away.**
- **5 51 Pegasi: Yellow G-type star; slightly larger than the Sun; 50 light-years away.**

STELLAR TARGETS FOR TPF SEARCHES:

- **1 Epsilon Indi A: Orange-red K-type star, about 75 percent the size of the Sun; 12 light-years away.**
- **2 Epsilon Eridani: Orange-red K-type star, about 85 percent the size of the Sun; 10.5 light-years away.**
- **3 Omicron[sup Eridani; Orange-red K-type star, about 90 percent of the Sun's size; 16 light-years away; in a triple system.**
- **4 Alpha Centauri B: Orange-red K-type star, about 90 percent the Sun's size; a member of the closest star system to Earth, about 4.3 light-years away.**
- **5 Tau Ceti: Orange-yellow G-type star, about 80 percent the Sun's size; 12 light-years away.**

HOW TO BE A GOOD PARENT STAR

DO remain at approximately the same luminosity for billions of years.

DO form out of a cloud with enough metal content to build up terrestrial planets.

DO stay in between galactic spiral arms for as long as possible.

DO NOT emit gigantic flares that singe the surfaces of nearby planets.

DO NOT form with companion stars that swoop in and out of your habitable zone.

Searching for Earth's History Among Earth-like Worlds[*]

By Lisa Kaltenegger

Mercury (San Francisco, Calif.), Winter 2007

It is only a matter of time before astronomers find an Earth-sized planet orbiting a distant star. We will naturally ask whether the planet is habitable or bears life. But will we be able to tell if there are bacteria, roaming dinosaurs, or even more advanced life?

While the first images of any putative exoplanet will be only visual smudges a single pixel in size, even a low-resolution spectral picture of the planet will be able to tell us a great deal. For examples: from spectra we should be able to infer whether the world, like Earth four billions years ago, was enveloped in a steamy, oxygenless atmosphere and covered completely by an ocean; or, in a Jurrasic-park-like epoch, a distant planet's atmosphere consists of about three-fourths nitrogen and one-fourth oxygen, with a small percentage of other gases like carbon dioxide and methane.

In the past twelve years, scientists have discovered more than two hundred large planets orbiting other stars, yet finding Earth-like planets is a fascinating and substantially more technologically demanding endeavor. And, once we finally do find them, what are clues to life's presence on them? Indeed, how do we need to design our instruments so we don't miss life's spectral signature?

CLUES FROM HOME

From their spectra, exoplanets can be screened for habitability. In a famous paper that appeared in *Nature* in 1993 (365, pp. 715–21), Carl Sagan and four colleagues analyzed a spectrum of Earth collected by the Galileo probe during its flyby of our planet on the way to Jupiter. They searched that spectrum for signatures of life and concluded that the large amount of O_2 and the simultaneous presence of CH_4 traces are suggestive of biological activity. Moreover, their detection of a widespread red-absorbing pigment with no likely mineral origin supports the hypothesis of biophotosynthesis. The results of this search for signs of life on our own world implies that we need to gather as much information as possible in order to understand how the atmospheres of possibly very different planets operate physically and chemically.

To date, all exoplanets have been studied indirectly—for example, by monitoring the way the host star wobbles as the planet's gravity tugs on it. Only four exoplanets have been detected directly, and all are massive Jupiter-sized worlds. The next generation of space-based missions, like NASMS Terrestrial Planet Finder (TPF) and the European Space Agency's Darwin mission, will be able to study directly nearby Earth-sized worlds and their atmospheres in the visible and infrared portions of the electromagnetic spectrum. Particular gases leave highly visible signatures in a planet's spectrum, like fingerprints or DNA markers. And by spotting those fingerprints, researchers can learn about an atmosphere's composition and even deduce the presence of clouds.

A BACKWARD VIEW FORWARD

Earth's past can inform us about other planets' present. Indeed, if we turn back the clock and create models of the particular spectral fingerprints for different epochs during Earth's life, we can compare those model "fingerprints" to what we find in exoplanets' spectra. This comparison should then permit us to identify a distant world's current evolutionary stage—to learn, in a sense, if there it is a time of methane bacteria or photosynthesis or dinosaurs.

For clues to how we might use this process, my colleagues Wesley Traub (NASA Jet Propulsion Laboratory) and Kenneth Jucks (Center for Astrophysics) and I have looked at our home planet. Past geologic records show that Earth's atmosphere has changed dramatically over the past four-and-a-half billion years—in part because of the developing life forms on our planet as discussed in detail by James Kasting (Pennsylvania State University) and David Catling (University of Washington) in their article in the 2003 Annual Review of Astronomy and Astrophysics. Looking for similar atmospheric compositions on other worlds, we will have indicators if that planet has life on it and also in what evolutionary stage it is. Further, major events in our own planet's history have changed the atmosphere, and we find an intriguing sequence of spectral fingerprints that reflects those major events on Earth.

Our world's atmosphere has evolved through six distinct epochs, each characterized by a particular mix of gases with its own distinct spectral fingerprint. The path has followed the transition from an atmosphere rich in CO_2 (epoch 0) to a CO_2/CH_4-rich one (epoch 3) to our present-day atmosphere (epoch 5). The absorption features of methane, which are undetectable in low-resolution in epochs 0 and 5, grow in strength from epoch 1 to epoch 2 and decrease from epoch 3 to epoch 4. The carbon dioxide decreases over Earth's age, and, while water is visible spectroscopically with comparable strength throughout Earth's history, the oxygen concentration grows from epoch 3 to epoch 5.

Putting Earth's evolution in the context of a single year, with 12:00 A.M. on 1 January as the time of Earth's formation, we find very different fingerprints about every two months. And to set this year in context, humans evolved approximately 59 seconds to midnight on December 31st.

CLOSER INSPECTIONS

Human activity has altered Earth's atmosphere by injecting carbon dioxide as well as gases like freon, which are used by our refrigerators. Can we identify the spectral fingerprints of those byproducts on other worlds—knowing, of course, that fridges, and, thus, probably take-out, are common phenomena? Although Earth-orbiting satellites and balloon experiments can measure these atmospheric changes here at home, detecting similar effects on a distant world is beyond even the capabilities of TPF and Darwin. It will take flotillas of future space-based infrared telescopes to be able to accomplish this. But even if we can not detect fridges on other planets, there is exciting information that we can extract from low-resolution spectra.

Let us take a closer look at the problem of observation of an Earth-like planet around a distant star. The Earth-Sun intensity ratio is about 10 million in the thermal infrared (~10 μm) and about 10 billion in the visible (~0.5 μm). That means for every one photon of Earth, 10 billion (visible) or 10 million (infrared) photons of the Sun arrive at the same time. For our observations of that exoplanet, we need to block out the light from the star to be able to collect the light from its relatively dim planet. We can do this physically by putting a mask in front of the telescope (like when you put a hand in front of your eyes to block out the light from a bright lamp) or by combining the light from a few smaller telescopes so that the starlight is suppressed and the planet's light can be detected. The interferometric systems suggested for Darwin and the TPF Interferometer (TPF-I) missions will operate in the thermal mid-IR (5–20 μn), where we can detect the heat emitted from a planet, and the corona-graph suggested for the TPF Coronagraph (TPF-C) will operate in the visible (0.5–1 μm), where we measure the starlight that is reflected by the planet.

Both the infrared and visible spectral regions contain atmospheric bio-indicators that can appear in the spectrum of a planet: CO_2, H_2O, O_3 CH_4, and N_2O in the thermal infrared, and H_2O, O_3, O_2, CH_4, and CO_2 in the visible to near-infrared. The presence or absence of these spectral features will indicate similarities or differences with the atmospheres of terrestrial planets.

Our search for signs of life is based on the assumption that extraterrestrial life shares fundamental characteristics with life on Earth, in that it requires liquid water as a solvent and has a carbon-based chemistry. We adopt this conservative approach to rule out false positives, completely aware that we will likely miss some habitable planets, where life has developed based on other chemistry or where the environment does not lead to a buildup of oxygen in the atmosphere.

And regardless of habitability, future space-based missions like Darwin and TPF will permit us to do comparative planetology on a wide variety of planets. The atmosphere and climatology of seemingly Earth-like planets can be outside our current understanding and will provide exciting opportunities to learn about and test our understanding of planets.

The next few decades offer us so much potential new knowledge and will broaden our views of planets and their developmental stages. Indeed, I believe we will come to know whether or not our habitable blue world is unique in the Universe and if there are other worlds like ours out there.

EPOCH 0

12 February: Putting Earth's evolution in the context of a single year, with 12:00 A.M. On 1 January as the time of Earth's formation, we find very different finger-prints about every two months. And to set this year in context, humans evolved approximately 59 seconds to midnight on 31 December.

At the beginning of Epoch 0, 3.9 billion years ago, the young Earth possessed a turbulent and steamy atmosphere composed mostly of nitrogen and carbon dioxide and hydrogen sulfide. The days were shorter, the Sun was dimmer, and more green-house gases were in the atmosphere to keep the surface from freezing completely. One ocean covered our entire planet and absorbed bombardment from incoming meteorites and comets. Carbon dioxide helped warm the planet because the infant Sun was a third less luminous than today. Although no fossils survive from this time period, isotropic signatures in Greenland rocks were left behind by life forms that may have been photosynthetic like modern-day plants.

EPOCH 1

17 March: By the time Epoch 1 began about 3.5 billion years ago, Earth's planetary landscape featured volcanic island chains poking out of the vast global ocean. The first life on Earth was anaerobic bacteria, which are bacteria that can live without oxy-gen. These bacteria pumped large amounts of methane into the planet's atmosphere, changing it in detectable ways. [...] If similar bacteria exist on another planet, future missions like TPF and Darwin may detect their fingerprint in the atmosphere— making the first signs of E.T. probably no radio or TV broadcasts but a methane spectral line from bacteria.

EPOCH 2

5 June: As Earth reached an age of 2.5 billion years, Epoch 2 began, and the atmosphere reached its maximum methane concentration. The dominant gases were nitrogen, carbon dioxide, and methane. Discussions are ongoing as to whether methane hazes during this epoch could have darkened the skies. Continental landmasses were beginning to form. Stromatolites, masses of blue green algae like the ones found in western Australia, began pumping large amounts of oxygen into the atmosphere. No oxygen was seen in the atmosphere yet as the crust still was being oxidized, and no residual oxygen could build up in the atmosphere. Big changes were about to happen.

EPOCH 3

16 July: 2.0 billion years ago at the start of Epoch 3, the temperature on Earth's surface was probably high, making conditions right for heat-loving methane bacteria to thrive. But a new species evolved that shifted the atmosphere's balance permanently—the first photosynthetic organisms. They produced oxygen, a highly reactive gas that cleared out much of the methane, while also suffocating the anaerobic bacteria that produced it. In doing so, the planet's atmosphere gained its first free oxygen. The fight between the oxygen and methane-producing bacteria was on and, luckily for us, the methane bacteria lost. The landscape was flat and damp. With volcanoes smoking in the distance, brilliantly colored pools of greenish-brown scum created a sheen on the stench-filled water. The oxygen revolution was fully underway.

EPOCH 4

13 October: 800 million years ago, Earth entered Epoch 4, with continuing increases in oxygen levels. This time period coincides with what is now known as the "Cambrian Explosion," an interval of time 550–500 million years ago during which most major animal groups first appear in the fossil records. Earth now would be covered with swamps, seas, and a few active volcanoes. The oceans would be teeming with life.

EPOCH 5

8 November: Finally, 300 million years ago in Epoch 5, life has moved from the oceans onto land. Earth's atmosphere has reached its current composition of primarily nitrogen and oxygen. This time marks the beginning of the Mesozoic era that will include the dinosaurs. The scenery looks like Jurassic Park on a Sunday afternoon with plants covering the surface of Earth. Note that from the beginning of this epoch on to the present, the vegetation on our planet could be detected by a remote observer with a very large telescope.

Where Are They?[*]

Why I Hope the Search for Extraterrestrial Life Finds Nothing

By Nick Bostrom
Technology Review, May/June 2008

People got very excited in 2004 when NASA's rover *Opportunity* discovered evidence that Mars had once been wet. Where there is water, there may be life. After more than 40 years of human exploration, culminating in the ongoing Mars Exploration Rover mission, scientists are planning still more missions to study the planet. The Phoenix, an interagency scientific probe led by the Lunar and Planetary Laboratory at the University of Arizona, is scheduled to land in late May on Mars's frigid northern arctic, where it will search for soils and ice that might be suitable for microbial life (see "Mission to Mars," November/December 2007). The next decade might see a Mars Sample Return mission, which would use robotic systems to collect samples of Martian rocks, soils, and atmosphere and return them to Earth. We could then analyze the samples to see if they contain any traces of life, whether extinct or still active.

Such a discovery would be of tremendous scientific significance. What could be more fascinating than discovering life that had evolved entirely independently of life here on Earth? Many people would also find it heartening to learn that we are not entirely alone in this vast, cold cosmos.

But I hope that our Mars probes discover nothing. It would be good news if we find Mars to be sterile. Dead rocks and lifeless sands would lift my spirit.

Conversely, if we discovered traces of some simple, extinct life-form—some bacteria, some algae—it would be bad news. If we found fossils of something more advanced, perhaps something that looked like the remnants of a trilobite or even the skeleton of a small mammal, it would be very bad news. The more complex the life-form we found, the more depressing the news would be. I would find it interesting, certainly—but a bad omen for the future of the human race.

How do I arrive at this conclusion? I begin by reflecting on a well-known fact. UFO spotters, Raëlian cultists, and self-certified alien abductees notwithstanding, humans have, to date, seen no sign of any extraterrestrial civilization. We have not received any visitors from space, nor have our radio telescopes detected any signals transmitted by any extraterrestrial civilization. The Search for Extra-Terrestrial Intelligence (SETI) has been going for nearly half a century, employing increasingly powerful telescopes and data–mining techniques; so far, it has consistently corroborated the null hypothesis. As best we have been able to determine, the night sky is empty and silent. The question "Where are they?" is thus at least as pertinent today as it was when the physicist Enrico Fermi first posed it during a lunch discussion with some of his colleagues at the Los Alamos National Laboratory back in 1950.

Here is another fact: the observable universe contains on the order of 100 billion galaxies, and there are on the order of 100 billion stars in our galaxy alone. In the last couple of decades, we have learned that many of these stars have planets circling them; several hundred such "exoplanets" have been discovered to date. Most of these are gigantic, since it is very difficult to detect smaller exoplanets using current methods. (In most cases, the planets cannot be directly observed. Their existence is inferred from their gravitational influence on their parent suns, which wobble slightly when pulled toward large orbiting planets, or from slight fluctuations in luminosity when the planets partially eclipse their suns.) We have every reason to believe that the observable universe contains vast numbers of solar systems, including many with planets that are Earth-like, at least in the sense of having masses and temperatures similar to those of our own orb. We also know that many of these solar systems are older than ours.

From these two facts it follows that the evolutionary path to life-forms capable of space colonization leads through a "Great Filter," which can be thought of as a probability barrier. (I borrow this term from Robin Hanson, an economist at George Mason University.) The filter consists of one or more evolutionary transitions or steps that must be traversed at great odds in order for an Earth-like planet to produce a civilization capable of exploring distant solar systems. You start with billions and billions of potential germination points for life, and you end up with a sum total of zero extraterrestrial civilizations that we can observe. The Great Filter must therefore be sufficiently powerful—which is to say, passing the critical points must be sufficiently improbable—that even with many billions of rolls of the dice, one ends up with nothing: no aliens, no spacecraft, no signals. At least, none that we can detect in our neck of the woods.

Now, just where might this Great Filter be located? There are two possibilities: It might be behind us, somewhere in our distant past. Or it might be ahead of us, somewhere in the decades, centuries, or millennia to come. Let us ponder these possibilities in turn.

If the filter is in our past, there must be some extremely improbable step in the sequence of events whereby an Earth-like planet gives rise to an intelligent species comparable in its technological sophistication to our contemporary human civilization. Some people seem to take the evolution of intelligent life on Earth for granted: a lengthy process, yes; complicated, sure; yet ultimately inevitable, or nearly so. But this view might well be completely mistaken. There is, at any rate, hardly any evidence to support it. Evolutionary biology, at the moment, does not enable us to calculate from first principles how probable or improbable the emergence of intelligent life on Earth was. Moreover, if we look back at our evolutionary history, we can identify a number of transitions any one of which could plausibly be the Great Filter.

For example, perhaps it is very improbable that even simple self-replicators should emerge on any Earth-like planet. Attempts to create life in the laboratory by mixing water with gases believed to have been present in the Earth's early atmosphere have failed to get much beyond the synthesis of a few simple amino acids. No instance of abiogenesis (the spontaneous emergence of life from nonlife) has ever been observed.

The oldest confirmed microfossils date from approximately 3.5 billion years ago, and there is tentative evidence that life might have existed a few hundred million years before that; but there is no evidence of life before 3.8 billion years ago. Life might have arisen considerably earlier than that without leaving any traces: there are very few preserved rock formations that old, and such as have survived have undergone major remolding over the eons. Nevertheless, several hundred million years elapsed between the formation of Earth and the appearance of the first known life-forms. The evidence is thus consistent with the hypothesis that the emergence of life required an extremely improbable set of coincidences, and that it took hundreds of millions of years of trial and error, of molecules and surface structures randomly interacting, before something capable of self-replication happened to appear by a stroke of astronomical luck. For aught we know, this first critical step could be a Great Filter.

Conclusively determining the probability of any given evolutionary development is difficult, since we cannot rerun the history of life multiple times. What we can do, however, is attempt to identify evolutionary transitions that are at least good candidates for being a Great Filter—transitions that are both extremely improbable and practically necessary for the emergence of intelligent technological civilization. One criterion for any likely candidate is that it should have occurred only once. Flight, sight, photosynthesis, and limbs have all evolved several times here on Earth and are thus ruled out. Another indication that an evolutionary step was very improbable is that it took a very long time to occur even after its prerequisites were in place. A long delay suggests that vastly many random recombinations occurred before one worked. Perhaps several improbable mutations had to occur all at once in order for an organism to leap from one local fitness peak to another: individually deleterious mutations might be fitness enhancing only when they occur together. (The evolution of *Homo sapiens* from our recent hominid ancestors, such as *Homo erectus*, happened rather quickly on the geological timescale, so these steps would be relatively weak candidates for a Great Filter.)

The original emergence of life appears to meet these two criteria. As far as we know, it might have occurred only once, and it might have taken hundreds of millions of years for it to happen even after the planet had cooled down enough for a wide range of organic molecules to be stable. Later evolutionary history offers additional possible Great Filters. For example, it took some 1.8 billion years for prokaryotes (the most basic type of single-celled organism) to evolve into eukaryotes (a more complex kind of cell with a membrane-enclosed nucleus). That is a long time, making this transition an excellent candidate. Others include the emergence of multicellular organisms and of sexual reproduction.

If the Great Filter is indeed behind us, meaning that the rise of intelligent life on any one planet is extremely improbable, then it follows that we are most likely the only technologically advanced civilization in our galaxy, or even in the entire observable universe. (The observable universe contains approximately 10^{22} stars. The universe might well extend infinitely far beyond the part that is observable by us, and it may contain infinitely many stars. If so, then it is virtually certain that an infinite number of intelligent extraterrestrial species exist, no matter how improbable their evolution on any given planet. However, cosmological theory implies that because the universe is expanding, any living creatures outside the observable universe are and will forever remain causally disconnected from us: they can never visit us, communicate with us, or be seen by us or our descendants.)

The other possibility is that the Great Filter is still ahead of us. This would mean that some great improbability prevents almost all civilizations at our current stage of technological development from progressing to the point where they engage in large-scale space colonization. For example, it might be that any sufficiently advanced civilization discovers some technology—perhaps some very powerful weapons technology—that causes its extinction.

I will return to this scenario shortly, but first I shall say a few words about another theoretical possibility: that extraterrestrials are out there in abundance but hidden from our view. I think that this is unlikely, because if extraterrestrials do exist in any numbers, at least one species would have already expanded throughout the galaxy, or beyond. Yet we have met no one.

Various schemes have been proposed for how intelligent species might colonize space. They might send out "manned" spaceships, which would establish colonies and "terraform" new planets, beginning with worlds in their own solar systems before moving on to more distant destinations. But much more likely, in my view, would be colonization by means of so-called von Neumann probes, named after the Hungarian-born prodigy John von Neumann, among whose many mathematical and scientific achievements was the concept of a "universal constructor," or a self-replicating machine. A von Neumann probe would be an unmanned self-replicating spacecraft, controlled by artificial intelligence and capable of interstellar travel. A probe would land on a planet (or a moon or asteroid), where it would mine raw materials to create multiple replicas of itself, perhaps using advanced forms of nanotechnology. In a scenario proposed by Frank Tipler in 1981, replicas would then be launched in various directions, setting in motion a multiplying colonization wave. Our galaxy is about 100,000 light-years across. If a probe were capable of traveling at one-tenth the speed of light, every planet in the galaxy could thus be colonized within a couple of million years (allowing some time for each probe that lands on a resource site to set up the necessary infrastructure and produce daughter probes). If travel speed were limited to 1 percent of light speed, colonization might take 20 million years instead. The exact numbers do not matter much, because the timescales are at any rate very short compared with the astronomical ones on which the evolution of intelligent life occurs.

If building a von Neumann probe seems very difficult—well, surely it is, but we are not talking about something we should begin work on today. Rather, we are considering what would be accomplished with some very advanced technology of the future. We might build von Neumann probes in centuries or millennia—intervals that are mere blips compared with the life span of a planet. Considering that space travel was science fiction a mere half-century ago, we should, I think, be extremely reluctant to proclaim something forever technologically infeasible unless it conflicts with some hard physical constraint. Our early space probes are already out there: Voyager 1, for example, is now at the edge of our solar system.

Even if an advanced technological civilization could spread throughout the galaxy in a relatively short period of time (and thereafter spread to neighboring galaxies), one might still wonder whether it would choose to do so. Perhaps it would prefer to stay at home and live in harmony with nature. However, a number of considerations make this explanation of the great silence less than plausible. First, we observe that life has here on Earth manifested a very strong tendency to spread wherever it can. It has populated every nook and cranny that can sustain it: east, west, north, and south; land, water, and air; desert, tropic, and arctic ice; underground rocks, hydrothermal vents, and radioactive-waste dumps; there are even living beings inside the bodies of other living beings. This empirical finding is of course entirely consonant with what one would expect on the basis of elementary evolutionary theory. Second, if we consider our own species in particular, we find that it has spread to every part of the planet, and we have even established a presence in space, at vast expense, with the International Space Station. Third, if an advanced civilization has the technology to go into space relatively cheaply, it has an obvious reason to do so: namely, that's where most of the resources are. Land, minerals, energy: all are abundant out there yet limited on any one home planet. These resources could be used to support a growing population and to construct giant temples or supercomputers or whatever structures a civilization values. Fourth, even if most advanced civilizations chose to remain nonexpansionist forever, it wouldn't make any difference as long as there was one other civilization that opted to launch the colonization process: that expansionary civilization would be the one whose probes, colonies, or descendants would fill the galaxy. It takes but one match to start a fire, only one expansionist civilization to begin colonizing the universe.

For all these reasons, it seems unlikely that the galaxy is teeming with intelligent beings that voluntarily confine themselves to their home planets. Now, it is possible to concoct scenarios in which the universe is swarming with advanced civilizations every one of which chooses to keep itself well hidden from our view. Maybe there is a secret society of advanced civilizations that know about us but have decided not to contact us until we're mature enough to be admitted into their club. Perhaps they're observing us as if we were animals in a zoo. I don't see how we can conclusively rule out this possibility. But I will set it aside in order to concentrate on what to me appear more plausible answers to Fermi's question.

The more disconcerting hypothesis is that the Great Filter consists in some destructive tendency common to virtually all sufficiently advanced technological civilizations. Throughout history, great civilizations on Earth have imploded—the Roman Empire, the Mayan civilization that once flourished in Central America, and many others. However, the kind of societal collapse that merely delays the eventual emergence of a space-colonizing civilization by a few hundred or a few thousand years would not explain why no such civilization has visited us from another planet. A thousand years may seem a long time to an individual, but in this context it's a sneeze. There are probably planets that are billions of years older than Earth. Any intelligent species on those planets would have had ample time to recover from repeated social or ecological collapses. Even if they failed a thousand times before they succeeded, they still could have arrived here hundreds of millions of years ago.

The Great Filter, then, would have to be something more dramatic than run-of-the mill societal collapse: it would have to be a terminal global cataclysm, an existential catastrophe. An existential risk is one that threatens to annihilate intelligent life or permanently and drastically curtail its potential for future development. In our own case, we can identify a number of potential existential risks: a nuclear war fought with arms stockpiles much larger than today's (perhaps resulting from future arms races); a genetically engineered superbug; environmental disaster; an asteroid impact; wars or terrorist acts committed with powerful future weapons; superintelligent general artificial intelligence with destructive goals; or high-energy physics experiments. These are just some of the existential risks that have been discussed in the literature, and considering that many of these have been proposed only in recent decades, it is plausible to assume that there are further existential risks we have not yet thought of.

The study of existential risks is an extremely important, albeit rather neglected, field of inquiry. But in order for an existential risk to constitute a plausible Great Filter, it must be of a kind that could destroy virtually any sufficiently advanced civilization. For instance, random natural disasters such as asteroid hits and supervolcanic eruptions are poor Great Filter candidates, because even if they destroyed a significant number of civilizations, we would expect some civilizations to get lucky; and some of these civilizations could then go on to colonize the universe. Perhaps the existential risks that are most likely to constitute a Great Filter are those that arise from technological discovery. It is not far-fetched to imagine some possible technology such that, first, virtually all sufficiently advanced civilizations eventually discover it, and second, its discovery leads almost universally to existential disaster.

So where is the Great Filter? Behind us, or not behind us?

If the Great Filter is ahead of us, we have still to confront it. If it is true that almost all intelligent species go extinct before they master the technology for space colonization, then we must expect that our own species will, too, since we have no reason to think that we will be any luckier than other species. If the Great Filter is ahead of us, we must relinquish all hope of ever colonizing the galaxy, and we must fear that our adventure will end soon—or, at any rate, prematurely. Therefore, we had better hope that the Great Filter is behind us.

What has all this got to do with finding life on Mars? Consider the implications of discovering that life had evolved independently on Mars (or some other planet in our solar system). That discovery would suggest that the emergence of life is not very improbable. If it happened independently twice here in our own backyard, it must surely have happened millions of times across the galaxy. This would mean that the Great Filter is less likely to be confronted during the early life of planets and therefore, for us, more likely still to come.

If we discovered some very simple life-forms on Mars, in its soil or under the ice at the polar caps, it would show that the Great Filter must come somewhere after that period in evolution. This would be disturbing, but we might still hope that the Great Filter was located in our past. If we discovered a more advanced life-form, such as some kind of multicellular organism, that would eliminate a much larger set of evolutionary transitions from consideration as the Great Filter. The effect would be to shift the probability more strongly against the hypothesis that the Great Filter is behind us. And if we discovered the fossils of some very complex life-form, such as a vertebrate-like creature, we would have to conclude that this hypothesis is very improbable indeed. It would be by far the worst news ever printed.

Yet most people reading about the discovery would be thrilled. They would not understand the implications. For if the Great Filter is not behind us, it is ahead of us. And that's a terrifying prospect.

So this is why I'm hoping that our space probes will discover dead rocks and lifeless sands on Mars, on Jupiter's moon Europa, and everywhere else our astronomers look. It would keep alive the hope of a great future for humanity.

Now, it might be thought an amazing coincidence if Earth were the only planet in the galaxy on which intelligent life evolved. If it happened here, the one planet we have studied closely, surely one would expect it to have happened on a lot of other planets in the galaxy—planets we have not yet had the chance to examine. This objection, however, rests on a fallacy: it overlooks what is known as an "observation selection effect." Whether intelligent life is common or rare, every observer is guaranteed to originate from a place where intelligent life did, in fact, arise. Since only the successes give rise to observers who can wonder about their existence, it would be a mistake to regard our planet as a randomly selected sample from all planets. (It would be closer to the mark to regard our planet as a random sample from the subset of planets that did engender intelligent life, this being a crude formulation of one of the saner ideas extractable from the motley ore referred to as the "anthropic principle.")

Since this point confuses many, it is worth expanding on it slightly. Consider two different hypotheses. One says that the evolution of intelligent life is a fairly straightforward process that happens on a significant fraction of all suitable planets. The other hypothesis says that the evolution of intelligent life is extremely complicated and happens perhaps on only one out of a million billion planets. To evaluate their plausibility in light of your evidence, you must ask yourself, "What do these hypotheses predict I should observe?" If you think about it, both hypotheses clearly predict that you should observe that your civilization originated in places where intelligent life evolved. All observers will share that observation, whether the evolution of intelligent life happened on a large or a small fraction of all planets. An observation-selection effect guarantees that whatever planet we call "ours" was a success story. And as long as the total number of planets in the universe is large enough to compensate for the low probability of any given one of them giving rise to intelligent life, it is not a surprise that a few success stories exist.

If—as I hope is the case—we are the only intelligent species that has ever evolved in our galaxy, and perhaps in the entire observable universe, it does not follow that our survival is not in danger. Nothing in the preceding reasoning precludes there being steps in the Great Filter both behind us and ahead of us. It might be extremely improbable both that intelligent life should arise on any given planet and that intelligent life, once evolved, should succeed in becoming advanced enough to colonize space.

But we would have some grounds for hope that all or most of the Great Filter is in our past if Mars is found to be barren. In that case, we may have a significant chance of one day growing into something greater than we are now.

In this scenario, the entire history of humankind to date is a mere instant compared with the eons that still lie before us. All the triumphs and tribulations of the millions of people who have walked the Earth since the ancient civilization of Mesopotamia would be like mere birth pangs in the delivery of a kind of life that hasn't yet begun. For surely it would be the height of naïveté to think that with the transformative technologies already in sight—genetics, nano-technology, and so on—and with thousands of millennia still ahead of us in which to perfect and apply these technologies and others of which we haven't yet conceived, human nature and the human condition will remain unchanged. Instead, if we survive and prosper, we will presumably develop some kind of posthuman existence.

None of this means that we ought to cancel our plans to have a closer look at Mars. If the Red Planet ever harbored life, we might as well find out about it. It might be bad news, but it would tell us something about our place in the universe, our future technological prospects, the existential risks confronting us, and the possibilities for human transformation—issues of considerable importance.

But in the absence of any such evidence, I conclude that the silence of the night sky is golden, and that in the search for extraterrestrial life, no news is good news.

Bibliography

Books

Belfiore, Michael. *Rocketeers: How a Visionary Band of Business Leaders, Engineers, and Pilots is Boldly Privatizing Space*. New York: Smithsonian Books, 2007.

Bilstein, Roger E. *Stages to Saturn: A Technological History of the Apollo/Saturn Launch Vehicles*. Washington, D.C.: Scientific and Technical Information Branch, National Aeronautics and Space Administration, 1980.

Brzezinski, Matthew. *Red Moon Rising*. New York: Times Books, 2007.

Burrough, Bryan. *Dragonfly: NASA and the Crisis Aboard MIR*. New York: HarperCollins, 1998.

Burrows, William E. *Exploring Space: Voyages in the Solar System and Beyond*. New York: Random House, 1990.

————— . *This New Ocean: The Story of the First Space Age*. New York: Random House, 1998.

Casoli, Fabienne & Thérèse Encrenaz. *The New Worlds: Extrasolar Planets*. New York: Springer, 2007.

Cernan, Eugene & Don Davis. *The Last Man on the Moon: Astronaut Eugene Cernan and America's Race in Space*. New York: St. Martin's Press, 1999.

Chaikin, Andrew. *A Man on the Moon*. New York: Viking, 1994.

Collins, Michael. *Carrying the Fire: An Astronaut's Journeys*. New York: Farrar, Straus and Giroux, 1974.

French, Francis & Colin Burgess. *Into That Silent Sea: Trailblazers of the Space Era, 1961-1965*. Lincoln, Neb.: University of Nebraska Press, 2007.

————— . *In the Shadow of the Moon: A Challenging Journey to Tranquility, 1965-1969*. Lincoln, Neb.: University of Nebraska Press, 2007.

Gray, Mike. *Angle of Attack: Harrison Storms and the Race to the Moon*. New York: W. W.

Norton, 1992.

Hansen, James R. *First Man: The Life of Neil A. Armstrong*. New York: Simon & Schuster, 2005.

Harford, James. *Korolev: How One Man Masterminded the Soviet Drive to Beat America to the Moon*. New York: Wiley, 1997.

Harvey, Brian. *China's Space Program: From Conception to Manned Spaceflight*. New York: Springer, 2004.

Heppenheimer, T. A. *Countdown: A History of Space Flight*. New York: John Wiley & Sons, 1997.

Kranz, Gene. *Failure is Not an Option: Mission Control from Mercury to Apollo 13 and Beyond*. New York: Simon & Schuster, 2000.

Linehan, Dan. *SpaceShipOne: An Illustrated History*. Minneapolis, Minn.: Zenith Press, 2008.

Lovell, Jim & Jeffrey Kluger. *Lost Moon: The Perilous Voyage of Apollo 13*. Boston: Houghton Mifflin, 1994.

Mullane, Mike. *Riding Rockets: The Outrageous Tales of a Space Shuttle Astronaut*. New York: Scribner, 2006.

Neufeld, Michael J. *Von Braun: Dreamer of Space, Engineer of War*. New York: Alfred A. Knopf, 2007.

Sagan, Carl. *Cosmos*. New York: Random House, 1980.

————. *Pale Blue Dot: A Vision of the Human Future in Space*. New York: Random House, 1994.

Shirley, Donna. *Managing Martians*. New York: Broadway Books, 1998.

Slayton, Donald K. Dekel. *Deke! U.S. Manned Space: From Mercury to the Shuttle*. New York: Forge, 1994.

Smith, Andrew. *Moondust: In Search of the Men Who Fell to Earth*. New York: Fourth Estate, 2005.

Squyres, Steven. *Roving Mars: Spirit, Opportunity, and the Exploration of the Red Planet*. New York: Hyperion, 2005.

Thompson, Milton O. *At the Edge of Space: The X-15 Flight Program*. Washington, D.C.: Smithsonian Institution Press, 1992.

Wolfe, Tom. *The Right Stuff*. New York: Farrar, Straus and Giroux, 1979.

Zubrin, Robert & Richard Wagner. *The Case for Mars: The Plan to Settle the Red Planet and Why We Must.* New York: The Free Press, 1996.

Web sites

Readers seeking additional information pertaining to manned and unmanned space exploration may wish to consult the following Web sites, all of which were operational as of this writing.

Great Images at NASA (GRIN)

grin.hq.nasa.gov/index.html

This Web site, maintained by the National Aeronautics and Space Administration (NASA), is a searchable collection of more than 1,000 photographs depicting various aspects of space exploration.

National Aeronautics and Space Administration (NASA)

www.nasa.gov

NASA has been charged with running America's space program since its establishment in July 1958. In addition to overseeing U.S. manned and unmanned space flight, the agency is also responsible for long-term civilian and military aerospace research. Its Web site provides text, images, and video that present not only an overview of NASA's history but also its current programs, including an outline for its Vision for Space Exploration, which calls for a return to the Moon in the next decade.

National Space Society (NSS)

www.nss.org

Boasting 50 chapters across the globe with an estimated 12,000 members, the National Space Society (NSS), an independent nonprofit organization, is dedicated to establishing a human presence in outer space. The society was formed in 1987, when the National Space Institute (founded in 1974) and the L5 Society (established in 1975) merged. This Web site provides an overview of the society's mission as well as information on *Ad Astra*, its award-winning magazine.

Russian Federal Space Agency

www.roscosmos.ru/index.asp?Lang=ENG

The Russian Federal Space Agency, also known as "Roskosmos" or by the initials RKA or RSA, is the successor to the former Soviet Union's space agency. This English-language version of its official Web site details its current efforts in space, including its involvement with the International Space Station and its long-running Soyuz capsule program.

Views of the Solar System

www.solarviews.com/eng/homepage.htm

This Web site, created by Calvin J. Hamilton, is one of the most comprehensive devoted to space exploration that is maintained by a private individual. Employing a multimedia approach, Hamilton uses photographs, scientific facts, text, graphics, and videos to provide visitors with a planet-by-planet understanding of our solar system.

Additional Periodical Articles with Abstracts

More information about the coming space age and related subjects can be found in the following articles. Readers interested in additional articles may consult the *Readers' Guide to Periodical Literature* and other H.W. Wilson publications.

NASA's Vision: An Outdated Business Model? Joan C. Horvath. *Ad Astra*, v. 18 pp30–32 Winter 2006.

As NASA plans to return to the Moon and go on to Mars, the writer wonders if mimicking the Apollo program's development will work in today's political and economic climate. She examines whether NASA's Vision for Space Exploration (VSE) is a workable technological and programmatic approach given the lessons learned during Apollo, offering assessments from a number of experts, including Roger Launius, chair of the Division of Space History at the Smithsonian's National Air and Space Museum, who believes that society is more risk averse today than during the Apollo era, and Donna Shirley, former manager of the JPL Mars Exploration program, who does not think a Moon mission will necessarily serve as an effective dress rehearsal for Mars. Horvath notes that the national will seems to be a vital factor in these missions, and that, for the VSE to succeed, it will be necessary to revive the optimism evoked by the pioneers of the Apollo program.

NASA: The Next 50 Years. Jeff Foust. *Ad Astra*, v. 19 pp24–25 Fall 2007.

NASA will continue to think of bigger and bolder missions despite suffering limitations in time, technology, and money, Foust predicts. The agency's current long-range plans are encapsulated in the Vision for Space Exploration (VSE), which seeks a manned mission to the Moon no later than 2020, to be followed, at an unspecified date, by manned missions to Mars and other destinations in the solar system. Assuming the 2020 objective can be met—no easy feat—NASA administrator Michael Griffin believes that a mission to Mars could be launched as soon as the late 2020s. For the time being, however, exploration of the outer solar system will be carried out by robotic spacecraft, and missions such as Galileo and Cassini have stimulated scientific interest in the giant planets and their retinues of moons.

Autonomy in Space: Current Capabilities and Future Challenges. Ari Jonsson, Robert A. Morris, and Liam Pedersen. *AI Magazine*, v. 28 pp27–42 Winter 2007.

In this article the authors provide an overview of the nature and role of autonomy in space exploration, particularly as it relates to artificial intelligence (AI) technologies. They analyze the range of autonomous behavior that is relevant and useful in space exploration and illustrate the range of possible behaviors by presenting four case studies in space-exploration systems, each differing from the others in their degree of autonomy. Three core requirements are defined for autonomous space systems, and the architectures for integrating capabilities into an autonomous system are described. The authors conclude with a discussion of the challenges that are faced currently in developing and deploying autonomy technologies in space.

Space Legends: Robert Zubrin. David Isaac. *The American Enterprise*, v. 15 pp20–23 December 2004.

Isaac presents an interview with Robert Zubrin, an advocate of manned exploration of Mars. Zubrin discusses why Mars should be explored; the influence of the 1960s Apollo missions on his career; his role as senior aerospace engineer for the Martin Marietta Company, in the mission to get humans to Mars; President George W. Bush's timetable for Mars exploration; the achievements of the Mars Society, of which he is president; the rationale for NASA's research of zero gravity; the future direction of space exploration; and the arguments for terraforming, or transforming another planet into one that is habitable by life from Earth.

In the Zone. Katherine J. Mack. *American Scientist*, v. 96 pp108–09 March/April 2008.

A new technique for detecting exoplanets was announced at a recent meeting of the American Astronomical Society, Mack reports. With current search methods, it is often difficult to detect exoplanets. However, Rosanne Di Stefano, an astronomer at the Harvard Smithsonian Center for Astrophysics, proposed a new kind of lensing search called mesolensing, which is optimized to detect planets circling in habitable zones and could revolutionize the way astronomers search for other worlds. The most exciting aspect of mesolensing, however, is its potential to detect planets suitable for life.

Portraits from Mars. Jim Bell. *Astronomy,* v. 35 pp64–69 January 2007.

Images of the surface of Mars taken by the Mars Exploration Rovers *Spirit* and *Opportunity* transform raw science into artistic landscapes, the author notes. Compared with the views of Mars from earlier missions, the rovers allow more time to be devoted to picture-taking, in addition to providing higher bandwidth and better camera resolution. These advantages allow for a more artistic scope while gathering all of the scientific and engineering information needed to run the missions. The images reveal the beauty, desolation, grandeur, and alien strangeness of the Martian landscape.

Space Probes Take Comet Close-ups. Francis Reddy. *Astronomy*, v. 35 p44E May 2007.

According to Reddy the first detailed views of comets have revealed that these bodies are more varied than was expected. The gas and dust that form a comet's coma and tail prevent astronomers from having a clear view of its nucleus, but a selection of images from the *Giotto, Deep Space 1*, and *Deep Impact* probes illustrate the great variety of cometary nuclei.

Managing Out of This World. Esther Rudis. *Conference Board Review*, v. 44 pp38–44 November/December 2007.

The Mars Exploration Rover (MER) story stands out as an illustration of how a senior management team had to continually adapt, rethink, and reinvent itself, Rudis explains. The aim of the MER project was to send two wheeled robots to Mars to explore the planet's surface. The way in which the project succeeded through the innovation of the management team is discussed.

Planet Hunter. Stephen Fraser. *Current Science*, v. 93 pp10–11 April 11, 2008.

Fraser examines the work of Sara Seager, an astronomer at the Massachusetts Institute of Technology (MIT), who is at the forefront of the search for exoplanets. Her goal is finding one that could be habitable to humans. From satellite images of Earth, Seager's team has discovered that the planet's cloud cover emits a repeating pattern of light. A similar pattern seen radiating from an exoplanet might signal the presence of a world very much like our own.

Feeling Out of This World. Richard Fisher. *The Engineer*, v. 293 p14 June 25–July 8, 2004.

Fisher reports that researchers at Cranfield and Leicester universities in the United Kingdom are working in collaboration with the European Space Agency (ESA) to develop lighter and more sensitive biosensors to allow future space probes to detect and identify evidence of life on other planets more precisely. They are currently working on an instrument called the Specific Molecular Identification of Life Experiment (SMILE), the transducers of which will be coated with biological receptors such as antibodies or enzymes. These biosensors will change their properties when in contact with an extraterrestrial molecule, such as an amino acid or a protein. SMILE is proposed for use on the ESA ExoMars rover mission, due to launch in 2009.

Starring Role. Niall Firth. *The Engineer*, v. 293 pp34–35 June 13–26, 2005.

Firth examines the career of Jim Benson, chief executive and founder of the private space technology company SpaceDev and a computer entrepreneur who made his name as founder of Compusearch in the mid-1980s. The future of private enterprise in space travel changed in October 2004, when *SpaceShipOne* became the first private spacecraft to reach low orbit. SpaceDev supplied the craft's hybrid thrusters, which were preferred over traditional thrusters as they were non-explosive and utilized a type of rubber as solid fuel and nitrous oxide as the gaseous oxidizer. SpaceDev has now agreed on a contract to launch micro-satellites for the U.S. Air Force Research Laboratory.

Next Foreign Chemical Site: The Moon. Joe Kamalick, *ICIS Chemical Business Americas*, v. 271 pp18-19 January 15–21, 2007.

James Garvin, chief scientist at the Goddard Space Flight Center, says that chemistry will play a vital role in NASA's ambitious plan to establish a permanently manned base on the Moon, Kamalick reports. According to Garvin, NASA will have to use chemistry both to understand what is available on the Moon and to exploit those resources. NASA's aim is to put in place chemical processes and equipment that generate enough oxygen and hydrogen to sustain human life, build structures with lunar soil, and make fuels on the Moon for lunar use and the eventual exploration of Mars. The author also discusses the processes that could be used to extract oxygen from lunar soil .

Entrepreneur Of The Year '07: Elon Musk. Max Chafkin. *Inc.*, v. 29 pp114–25 December 2007.

Chafkin profiles Elon Musk, co-founder of PayPal and Zip2, the current CEO, majority owner, and head rocket designer at SpaceX, an El Segundo, California-based aerospace start-up that aims to start transporting astronauts to and from the International Space Station by 2011. Musk is also chairman and controlling shareholder of two ambitious start-ups: electric-car manufacturer Tesla Motors and solar-panel installer SolarCity.

To Know is Worth the Expense. Whitney Howell. *Mercury*, v. 33 p46 July/August 2004.

The writer believes that space exploration should not be contingent on expense. The question of cost dominates debates about space exploration, particularly as technological innovations yield further-reaching capabilities. However, for mankind, exploration represents a biological and moral imperative, and the solar system is the open field in which humanity may fulfill its most basic desire to seek and learn. Thus, President Bush's new Vision for Space Exploration, which, according to NASA, pursues compelling science and inspires the next generation of explorers, should not be subjected to excessive budgeting.

Beautiful Stranger: Photographs of Saturn Taken by the Cassini Spacecraft. Bill Douthitt. *National Geographic*, v. 210 pp38–57 December 2006.

The ambitious Cassini-Huygens mission is allowing scientists to closely examine Saturn, which contains clues about the formation of the solar system and the origins of life, Douthitt notes. The *Cassini* spacecraft left Earth in 1997 and began its four-year exploration of Saturn on June 30, 2004. On December 25, 2004, it launched the probe *Huygens*, which dived into the smoggy atmosphere of Saturn's most tantalizing moon, Titan. Titan's gentle winds contain a rich blend of organic molecules, some of which bear similarities to the compounds that yielded the raw material for life on Earth. One of the most surprising discoveries of the mission was made when the spacecraft detected simple carbon compounds around the south polar region of another moon, Enceladus. Life could be concealed in pockets of warm water beneath the ice. *Cassini's* mission may be extended, but scientists are already envisioning future space probes capable of looking for life on Enceladus and studying the origins of life on Titus. Photographs from the spacecraft are also presented.

Fly Me to the Moon. Kathy Finn. *New Orleans Magazine*, v. 42 pp38–39 April 2008.

In 2007 the Boeing Company of Huntsville, Alabama, announced that it will build the upper stage of a new launch vehicle named *Ares 1* at the Michoud Assembly Facility of the National Aeronautics and Space Administration (NASA) in eastern New Orleans, Finn reports. The launch vehicle is the centerpiece of NASA's Constellation program, which aims to send people to the Moon by 2020. In addition Lockheed Martin Corp., which for decades has built the space shuttle's external fuel tanks at the Michoud plant, won a contract in 2006 to build the Orion Crew Exploration Vehicle, which the Ares rocket will carry into space. NASA's Sheila Cloud, who is directing the transition from the shuttle program to Constellation at Michoud, notes that the Ares 1 and Orion projects together have the potential to generate several hundred jobs in the local area.

Hello Earth, Messenger Calling. Jeff Hecht. *New Scientist*, v. 189 p16 January 14, 2006.

Hecht notes that NASA engineers have exchanged laser pulses with the *Messenger* spacecraft, 24 million kilometers from Earth. Space probes usually communicate with Earth using microwaves, but these are not as directional as lasers, so the signal spreads more and the data rate is limited. No data were transmitted in the laser experiment, but precise measurements of the probe's distance and speed were obtained, and the potential of lasers for interplanetary communication was demonstrated.

Bring on the Little Green Men. Zoë Williams. *New Statesman*, v. 135 pp30–31 March 13, 2006.

In this article Williams presents excerpts from an interview with Charles Moss Duke Jr., who became the tenth man to walk on the Moon during the Apollo 16 mission in 1972. Among the topics discussed are the difference between attitudes toward space exploration in the 1960s and today; reasons for the current lack of enthusiasm for space exploration; financial support for space exploration; China's space program; and the role that private investors could play in developing space travel.

A Waste of Space. Robin McKie. *New Statesman*, v. 137 pp32–34 April 7, 2008.

As NASA celebrates its 50th birthday much will be made of the agency's glory days, but behind the revelry is a sense of unease, McKie observes. Although it has recorded great achievements, NASA is also plagued by major political and financial problems. Questions arise as to whether one of the world's most lavishly funded scientific organization has justified its $16 billion annual budget, given what it has achieved for science and what it plans to accomplish in the future. Overall, NASA has the expertise to explore space, but to exploit space the agency will need to liberate itself from the shackles of its past.

Space: More Than Meets the Eye. *The OECD Observer*, no. 263 pp16–18 October 2007.

Although the space industry is relatively small, its economic and strategic importance is immense, the author remarks. In addition to Russia, the United States, and Europe—and more recently Japan, China, and India—there has been an upsurge in interest in space from such countries as South Korea, Turkey, and Nigeria. A number of states now have stakes in satellites that are currently in orbit and which carry scientific, broadcasting, and observation technology. The writer likewise discusses the strategic significance and ultimate value of the industry.

Extending the Search for Extrasolar Planets. Katie Walker. *Science & Technology Review*, pp4–10 March/April 2008.

Until now, scientists have had no way to study the majority of extrasolar planets or their atmospheres, but that is set to change when the Gemini Planet Imager (GPI) goes online at the Gemini South telescope in Chile in 2010, Walker reports. The imager's main goal will be to detect more planets outside the Solar System, providing important new information about how planets form and solar systems evolve. The most exciting component on GPI is a spectrograph, which will calculate the infrared light emitted by a planet's atmosphere. With the spectrograph the imager will identify the chemical makeup of the atmosphere. Scientists can utilize the atmospheric data to make conclusions about a planet's temperature, pressure, and gravity. Lawrence Livermore scientists form part of an international collaboration developing GPI, with Livermore astrophysicist Bruce Mackintosh heading the design team and engineer Dave Palmer serving as project manager.

Study Suggests No Dearth of Earths. Ron Cowen. *Science News*, v. 173 pp67–68 February 2, 2008.

New research suggests that Earth-like planets are orbiting or forming around many Sun-like stars, Cowen notes. In the February 1 *Astrophysical Research Letters*, Meyers and others describe the results of infrared survey of some 300 stars with similar masses to that of the Sun. The researchers found that 10 percent of the stars examined had an infrared wavelength that indicated the presence of dust, which suggests that terrestrial planets may be forming around the stars. Meyers and others estimate that up to 62 percent of the star systems surveyed could potentially have the materials and liquid water necessary for the formation of rocky planets orbiting in their inner regions. Meyer notes that actual proof of terrestrial exoplanets awaits next year's launch of the Kepler mission.

More Early Clues for Life at Home, Out There. Ron Cowen. *Science News,* v. 173 p181 March 22, 2008.

Astronomers have been uncovering clues to the origins of life both on Earth and, potentially, on extrasolar planets. In the March 20 issue of *Nature*, Swain and colleagues confirm previous suggestions that an extrasolar planet some 63 light-years from Earth has water vapor in its atmosphere and also report that the atmosphere contains methane—the first time any carbon-bearing compound has been detected in the atmosphere of an extrasolar planet. Moreover, in a forthcoming issue of *Meteorites & Planetary Science*, Martins and colleagues report findings that suggest that meteorites delivered much larger amounts of amino acids to the early Earth than had been previously assumed.

China's Great Leap Upward. James Oberg. *Scientific American*, v. 289 pp76–83 October 2003.

Oberg discusses the history and ambitions of China's space flight program. This fall China plans to become just the third nation to launch a manned space flight, which it hopes will stimulate advances in the country's aerospace industry and gain international prestige. China's space program dates back to the launch of the country's first satellite in 1970 and planning for the manned mission has been underway since President Jiang Zemin gave the go-ahead for the project in the early 1990s. Oberg also examines the Shenzhou craft that will carry the country's first astronauts.

Space Suit Redux. Lisa Scanlon. *Technology Review*, v. 108 pM9 June 2005.

An old design for a spacesuit could be the inspiration for the suit of the future, Scanlon reports. Researchers are reexamining a design for an elastic suit conceived by physiologist Paul Webb during the 1960s. The suit is basically a very tight leotard that applies pressure to the skin. NASA did not follow up on the concept because the suit proved too difficult to put on and remove; instead, it opted for the now familiar space suit that envelops the body with a balloon of pressurized oxygen. However, this design is impractical for NASA's long-term aim of human exploration of Mars. According to Dava Newman, a professor of aeronautics and astronautics, Webb's design is better suited for Mars, and with advances in materials, its time may have come. In a project sponsored by the NASA Institute for Advanced Concepts, Newman and her students made a prototype pant leg that melds the pressurized-gas and tight-fabric techniques. She is now hoping to make a working prototype of a spacesuit leg by the end of the summer, although the researchers warn that an operational suit may be decades off.

Meet the World's First Female Private Space Explorer. Laura Grover. *USA Today*, v. 135 pp56–58 January 2007.

Grover profiles Anousheh Ansari, who became the world's first female private space explorer when she worked as a primary care member on the *Soyuz TMA-9* for a recent eight-day mission aboard the International Space Station. As a pioneering space ambassador, Ansari hopes to build widespread awareness of and enthusiasm for space exploration, and to assist in promoting peace and understanding between nations. Ansari and Peter Diamandis, chairman and founder of the X Prize Foundation, share the long-term objective of making space travel safe, affordable, and accessible to everyone through the creation of a viable personal space-flight industry. Furthermore, they envision a future that is critically dependent on opening up the space frontier and utilizing its resources.

Guiding the Path to Mars. Alex Markels. *U.S. News & World Report*, v. 141 pp90–91 October 30, 2006.

Charles Elachi has lead NASA's Jet Propulsion Laboratories in Pasadena, California, to conquer Mars and the moons of Saturn, Markels reports. When Elachi was promoted to become director of the facility almost six years ago, morale was poor among the engineers and scientists there after the failure of the $327 million Mars Climate Orbiter and the $120 million Mars Polar Lander. Elachi, an unflappable Lebanese-born physicist, declined to point fingers, however. He turned down the Mars project manager's offer to resign and instead rallied the team to work on the Mars Exploration Rover, a project that included two spacecraft, *Spirit* and *Opportunity*. That mission launched on time, and only 17 percent over budget, in June 2003, and, after several tense days during which a computer glitch was smoothed out, the rovers started an exploration that quickly surpassed all expectations and continues to this day.

Index

Aldrin, Buzz, 83
Allen, Paul, 86
Anderson, John D., 124
Anderson, Reda, 73
Ansari, Anousheh, 73
Antonellis, Robert, 47–53
Appenzeller, Tim, 127–128
Armstrong, Neil, 19, 83

Ballard, Robert, 84
Barghoorn, Tyler, 107
Bennett, Steven, 80–81
Benson, Jim, 54
Bezos, Jeff, 66, 70–71, 79, 82
Bigelow, Robert, 66, 68–70, 74
Bodenheimer, Peter, 147
Boles, Walter Wesley, 22
Bostrom, Nick, 163–170
Bracher, Katherine, 93–94
Branson, Richard, 70–71, 74, 80
Brasier, Martin, 109
Brown, Blaine, 16
Brown, David, 104–105
Brown, Peter J., 56
Burrows, Adam, 139
Bush, George H.W., 86
Bush, George W., 12, 20, 29–30
Butler, Paul, 139, 144

Callas, John, 102
Carmack, John, 69, 78
Carmichael, Mary, 54–55
Cernan, Gene, 18
Charbonneau, David, 143
Chen Long, 49
Christian, Eric, 127
Clark, Larry, 24–25
Cochran, William, 139–140
Cray, Dan, 129–131

David, Leonard, 35–37, 77–81
Davidian, Ken, 75–76
Democritus, 144
Dingell, Charles, 5–11
Downs, Hugh, 65

Farrand, Bill, 95–100
Fedo, Christopher, 109–110
Fermi, Enrico, 164, 168
Fischer, Debra, 143, 146
Fishman, Charles, 12–18
Frail, Dale, 138
Friedman, Louis, 29–30, 127

Garvin, Jim, 131
Gimarc, Alex, 87
Goddard, Robert, 71
Godwin, Richard, 76
Goldin, Daniel, 106
Gray, David, 139
Greene, Kevin, 72
Griffin, Michael, 17–18, 26–28, 30, 38, 40, 58, 75
Grotzinger, John, 110
Gugliotta, Guy, 19–25, 38–41

Hagt, Eric, 39, 59
Hatfield, Skip, 13
Hawking, Stephen, 26
Herschel, William, 123
Huang Chunping, 49
Hubbard, G. Scott, 29–30

Johnson-Freese, Joan, 36, 38–41, 58
Johns, Bill, 14
Johns, William A., 5–11
Jones, Jeremy, 119–120

Kaltenegger, Lisa, 158–162
Kennedy, John. F., 67

Kinsella, Gary, 125
Krukin, Jeff, 75
Kulcinski, Gerald, 55

Larson, William, 23
Launius, Roger, 37
Le Verrier, Urbain-Jean-Joseph, 124
Lemmon, Mark, 104–105
Lewis, James, 36
Liu, Melinda, 54–55
Logsdon, John M., 30, 39
Lovis, Christophe, 145
Lowe, Donald, 110
Lu, Edward, 16
Lunine, Jonathan, 143

Machin, Koki, 14–15
Mangu-Ward, Katherine, 68–76
Marburger, Jack, 26
Marcy, Geoff, 139, 142, 144–150
Mari, Christopher, 137–141
Mayor, Michael G.E., 137–141, 144
McKay, David, 106, 108–109
Milstein, Michael, 83–85
Minkel, JR, 142–143
Morin, Lee, 16
Murray, William S. III, 47–53
Musk, Elon, 70, 78

Needell, Allan, 84

O'Leary, Beth Laura, 84
O'Neill, Gerard K., 65
Osborne, David, 67
Ouyang Ziyuan, 49

Pacini, Franco, 139
Persaud, Rocky, 70
Pomerantz, William, 83, 85
Price, Larry, 13

Queloz, Didier, 137–140, 144

Red, Michael, 16–17

Reynolds, Glenn Harlan, 65–67
Richardson, Bill, 79–80
Rickman, Douglas, 20
Rohrabacher, Dana, 41
Rosing, Minik, 110
Ross, Amy, 23
Rufai, Ahmed, 57, 59
Rusch, Roger, 57
Rutan, Burt, 66, 86

Sagan, Carl, 93, 107, 158
Saunders, Phillip C., 42–46
Schmitt, Harrison, 21
Schopf, J. William, 106–108
Schwartz, John, 29–30
Semeniuk, Ivan, 101–103
Sena, Alberto, 16
Shen Dingli, 56
Smith, Peter, 105, 130
Spennemann, Dirk H. R.,83–84
Spilker, Linda, 129–130
Squyres, Steve, 95, 102
Stafford, Tom, 18
Steele, Andrew, 108–112
Stone, Ed, 128

Taylor, Lawrence, 21–24, 55
Theisinger, Pete, 102
Tice, Michael, 110
Toporski, Jan, 112
Traub, Wesley, 159
Tsien Hsue-shen, 36–37, 46
Tumlinson, Rick, 74
Turnbull, Margaret, 151–157
Turyshev, Slava, 124–126
Tyler, Stanley, 107
Tyson, Neil deGrasse, 121–126

Vick, Charles, 85
Vogt, Steve, 149
Von Neumann, John, 166–167

White, Julie Kramer, 5–11
Whitehorn, Will, 74

Whitehouse, Martin, 110
Whittaker, William "Red", 85
Wingo, Dennis, 86–88
Wolszczan, Aleksander, 138

Yang Liwei, 39–40
Yardley, Jim, 56–59
Ye Zili, 54

Zhang Houying, 47
Zhang Oinwei, 40
Zhang Qingwei, 49–50
Zimmer, Carl, 106–112

"I have spent my fair share of time in the waiting room, that place of sitting and waiting for it to be my turn. Anyone who has waited for the next step, the fulfilled dream, or the break of the glass ceiling will benefit by reading this book. Dr. Day gives us practical steps not just to wait but work, until it is our turn."

Aaron Duvall, teaching pastor at Victory Highway Wesleyan Church in Painted Post, New York, and former director of spiritual formation at Ohio Christian University

"Tired of clichés about counting your blessings? Then you'll love this book! *It's Not Your Turn* throws out those tired, guilt-inducing admonitions to just put on a happy face for Jesus and shows us the surprising, practical—and even fun!—way to wait in God's upside-down kingdom. My soul felt lighter just reading this book—and yours will too!"

Sheila Wray Gregoire, blogger at ToLoveHonorandVacuum.com and author of *The Great Sex Rescue*

"*It's Not Your Turn* is transparent, relevant, intelligent, and loaded with profound spiritual insights learned through real-life experiences. Heather Thompson Day is the best friend we all need to tell us the truth with grace, understanding, and love."

Jory Micah, blogger and feminist theologian

"In *It's Not Your Turn*, Heather reminds us that the process is often more important than the outcome. This book is full of wisdom, faith, research, humor, and practical advice on what to do until it is your turn."

Christine Caine, founder of A21 and Propel Women

"While you are waiting, grow. While you are waiting, learn. While you are waiting, listen. It's your turn to become who you've always wanted to be."

Annie F. Downs, author of *That Sounds Fun* and host of the *That Sounds Fun Podcast*, from the foreword

"What God does in you as you wait is more important than what you are waiting for. This is the essence of what Heather Thompson Day is leading us to in this needed book. With refreshing vulnerability, wit, and poignant insights, she helps us navigate the frustrations that come through the soul-wearying habit of comparison."

Rich Villodas, pastor of New Life Fellowship Church and author of *The Deeply Formed Life*

"What do you do when everyone around you seems to be winning the race for success, perfection, and progress while it feels like your prayers are unanswered and your life is stuck? In *It's Not Your Turn*, Dr. Heather Thompson Day offers much-needed perspective. Drawing on biblical insights, current research, and poignant personal stories, she reminds us that God's activity in our lives is not limited to society's definition of or timetables for success. Furthermore, she contends that our waiting time is not wasting time but can be a God-infused opportunity for growth, transformation, and purpose. Heather has given us a valuable resource for every chapter in our lives."

Jo Saxton, leadership coach and author of *Ready to Rise*

"*It's Not Your Turn* is the perfect resource on what to do while we wait. What should we do with our time while waiting? Heather gives us the tools!"

Michelle Williams, actress and singer, former member of Destiny's Child

"Heather is real, transparent, and beautifully eloquent at reminding us of the importance of trusting that all things—especially God's timing—work together for our good. If you are in a season of waiting and uncertainty and struggling to see the bigger purpose of it all, it is most definitely your turn to read this book!"

Mandy Hale, author of *You Are Enough*

"Heather Thompson Day writes in a way that sinks deep into your spirit. Her honest self-assessments help reveal the places where we all get caught up in jealousy, competition, and envy when it comes to the life God has given us in contrast to the life God has given others. A timely read for anyone struggling to move beyond comparison to a patient, non-anxious posture where we wait for the divine goodness of 'our turn.'"

Casey Tygrett, spiritual director and author of *As I Recall: Discovering the Place of Memories in Our Spiritual Life*

"This book is wise, hilarious, vulnerable, practical, and utterly delightful. Heather winsomely unpacks insights from psychology, biology, her experience, and her faith to yank us from our visions of glory, prisons of despair, or tantrums of envy, and helps us take concrete steps to make life good and beautiful again. *It's Not Your Turn* is a wonderful book that serves up guidance and healing for the striving-yet-discouraged soul."

James Choung, author of *True Story* and *Real Life*

it's
not
your
turn

what to do while you're waiting for your

BREAKTHROUGH

Heather Thompson Day

Foreword by Annie F. Downs

An imprint of InterVarsity Press
Downers Grove, Illinois

InterVarsity Press
P.O. Box 1400, Downers Grove, IL 60515-1426
ivpress.com
email@ivpress.com

InterVarsity Press® is the book-publishing division of InterVarsity Christian Fellowship/USA®, a movement of students and faculty active on campus at hundreds of universities, colleges, and schools of nursing in the United States of America, and a member movement of the International Fellowship of Evangelical Students. For information about local and regional activities, visit intervarsity.org.

All Scripture quotations, unless otherwise indicated, are taken from The Holy Bible, New International Version®, NIV®. Copyright © 1973, 1978, 1984, 2011 by Biblica, Inc.™ Used by permission of Zondervan. All rights reserved worldwide. www.zondervan.com. The "NIV" and "New International Version" are trademarks registered in the United States Patent and Trademark Office by Biblica, Inc.™

While any stories in this book are true, some names and identifying information may have been changed to protect the privacy of individuals.

The author is represented by MacGregor & Luedeke Literary, Inc.

The publisher cannot verify the accuracy or functionality of website URLs used in this book beyond the date of publication.

Cover design and image composite: Faceout Studio
Interior design: Jeanna Wiggins
Images: page of emoticons: © Aguni / Shutterstock Images
　　　 traffic light: © Mike Kemp / Getty Images

ISBN 978-0-8308-4776-1 (print)
ISBN 978-0-8308-4777-8 (digital)

Printed in the United States of America ♾

InterVarsity Press is committed to ecological stewardship and to the conservation of natural resources in all our operations. This book was printed using sustainably sourced paper.

Library of Congress Cataloging-in-Publication Data
Names: Day, Heather Thompson, author.
Title: It's not your turn : what to do while you're waiting for your
　　breakthrough / Heather Thompson Day.
Description: Downers Grove, IL : InterVarsity Press, [2021] | Includes
　　bibliographical references.
Identifiers: LCCN 2021002279 (print) | LCCN 2021002280 (ebook) | ISBN
　　9780830847761 (print) | ISBN 9780830847778 (digital)
Subjects: LCSH: Dreams—Religious aspects—Christianity. | Expectation
　　(Psychology)—Religious aspects—Christianity. | Waiting (Philosophy)
Classification: LCC BR115.D74 D39 2021 (print) | LCC BR115.D74 (ebook) |
　　DDC 248.2/9—dc23
LC record available at https://lccn.loc.gov/2021002279
LC ebook record available at https://lccn.loc.gov/2021002280

P　25　24　23　22　21　20　19　18　17　16　15　14　13　12　11　10　9　8　7　6　5　4　3　2

Y　37　36　35　34　33　32　31　30　29　28　27　26　25　24　23　22　21

This book is dedicated to my husband, Seth Day,
and our three children, London, Hudson, and Sawyer.
Every turn I made led me to you.

Contents

Foreword by Annie F. Downs   1

1 It's Not Your Turn   3

2 It's Your Turn to Wait   15

3 It's Your Turn to Say It Out Loud   28

4 It's Your Turn to See   49

5 It's Your Turn to Think Small   67

6 It's Your Turn to Set the Goal   82

7 It's Your Turn to Network   101

8 It's Your Turn to Take a
Second Look at Power   117

9 It's Your Turn to Find Community   135

10 It's Your Turn to Re-envision God   150

11 It's Your Turn to Move on Maybe   168

12 It's Your Turn to Make Your Move   182

Acknowledgments   195

Notes   197

Foreword

Annie F. Downs

M Y COUNSELOR GAVE ME the strangest advice a few months ago. Almost out of the blue, she said, "You should start waiting in the longest lines you can find." She meant everywhere—at the grocery store, getting my car's emissions checked, ordering dinner last in our group of friends. I hadn't even been talking about patience or anything; I had been talking about my life in general. But she wanted me to wait more.

I do not want to wait more. I want to wait less. I look around my life and there seem to be lots of places where I see others get to go first and I have to wait my turn. It's frustrating. But there is also something really important about knowing when it is my turn and when it isn't.

My counselor wanted me to practice waiting my turn in what I could control since I'm getting ample practice in waiting in ways I can't control. So even when I'm running short on time, I pick a parking spot a little farther away (Get those steps in, am I right?). And even when I'm exhausted after work, I pick the longest line at the grocery store—the one behind that woman with all the coupons and more tiny things in her basket than one would have thought humanly possible.

And while I am waiting my turn, I am not wasting my time. I keep a book on my phone, and I read. No social media, no texting—I just read books. Books that tell me where my spot is on this planet and where my spot isn't. Books that tell me what is my work to do and what isn't. Books that stretch me in my relationships with humans and my relationship with God. Books that tell me what to do while I'm waiting for my turn—for groceries, for my car to be cleaned, for my prayers to be answered, for my dreams to come true.

Books like this one.

While you are waiting, grow. While you are waiting, learn. While you are waiting, listen. It's your turn to become who you've always wanted to be.

It's Not Your Turn

Much of the war against the devil
is about whether you'll quit.

BETH MOORE

I N THE BEGINNING, I had to fake it: the happiness, the peace, the congratulations. It all felt heavy to carry over the gap of where I currently was and where I wanted to be. I was a few years into my PhD and couldn't find a job. I applied to what felt like every higher education institution with an opening in my field and kept getting rejected. At some point, it started to feel personal.

For my daughter's first birthday, we planned a huge party. My little girl was turning one, and I wanted to celebrate her. I went to the store to buy food. I had just paid for my groceries when I realized I forgot to get paper party plates. As I handed my card to the woman on the register, I watched as the gap between me with all my education and her with this minimum wage job evaporated.

"Ma'am," she said. "Your card is declined." My face got hot. Paper plates are $2.50. Y'all. I did not have $2.50.

How in the world did this happen to me? I had a husband. I had a house. I had a daughter. I had nearly three degrees. *But I didn't have $2.50?* I was mortified. My husband and I got into the car with our groceries and drove home in complete silence. I cried myself to sleep that night. I felt like a failure. I remember emailing God a letter (there are actual sites that allow you to do that) and while, of course, I knew this email wasn't going to God's inbox, it felt therapeutic to hit send on all my grievances. *I thought God opened doors and windows? I thought God owned the cattle on a thousand hills? I thought God answered prayers? Where was my testimony?*

It felt like God had played me. I had done everything right. I focused on school. I excelled in my teaching. Yet, here I had nothing to show for it. At this same time, one of my best friends Jewel, called me. She had just been hired by NASA as a recruiter for their minority student program. I couldn't get a job teaching at a community college, and Jewel was now employed by NASA.

"I am so happy for you," I legit choked.

And it's not that I wasn't happy for her, I was. I was just also so deeply sad for myself. That was the moment I learned a lesson in my life that I've repeated to myself a hundred times since: *Heather, it's not your turn.* Sometimes, you show up to someone else's party. Sometimes you force yourself to clap when you really want to cry because emotions aren't always singular. You are allowed to feel sad for yourself while also being happy for what is happening to someone else. I clapped for Jewel because she deserved it. It wasn't my turn, but it was hers. And I had to be the friend she needed.

In today's culture, it's a race to the top of the ladder. According to Pew Research, millennials are the most educated generation.[1] No one does comparison quite like millennials. We have apps for everything, and yet Yale University found people are happier and healthier the less time they spend online.[2] We are the generation of hashtags and filters. Everything is created to project an image of who we want to be—which is never as we actually are. We try our hardest to be witty in 140 characters or less. We post photos of our nights out, and the scene is always way more intriguing than the night really was.

Once I went to the beach with a friend. She experienced nausea from her early pregnancy. She complained the entire time and never got in the water. Within an hour, she asked to leave. That night, she posted the two photos we took with the caption: "Fun in the sun." That was one of the first times I realized we have totally curated online lives that are almost nothing like what we live. We get dozens of comments, hundreds of likes, and it fuels our need to continue with the charade. Anything for a hit. If only our real lives felt as successful as our cropped ones. I can't think of a single millennial friend who hasn't had some type of struggle with either anxiety or depression. Not a single one. Which means though we may feel alone, we aren't. There are probably thousands, if not millions, of us sitting right now feeling as though it's never going to be our turn.

I think some of this is the reason we often feel unfulfilled. We want everyone around us to believe we have it all together—and we don't. We fear everyone else is living the lives they post and we are the only imposters. And so, the race is on. The race to perfection. The race to instant success and gratification. The race toward fame, promotion, and adoration. But what if life isn't meant to be raced through? What if it's meant to be lived?

My mentor is José Rojas. He was a spiritual adviser for two US presidents. On one of our first phone conversations, he said to me, "What if you'll actually get to where you want to be quicker by slowing down?" I didn't get it, but now I do. What if by rushing through the process, we end up as a rushed product? Jesus, who entered the world as a baby and would need to die as a man, tells us all we need to know about the heavenly value of process. *We* value the product. God values the entire process. What if by slowing down, you get to where you're going faster?

It's Hard to Keep Up

We are always in competition with one another because we have constant access to each other. There was a time when you only competed with your neighbors over Christmas lights and tacky lawn ornaments. Now you can't pee without seeing how much better than you 250 of your "closest" friends are. Newsfeeds are filled with all the awesome philanthropy, money, and stardom your old college roommates have found. Then we look at our own lives and we feel like crap and so we talk crap. I can't tell you how many group chats I've exited. Friends screenshotting people's posts to poke holes where we can. It makes us feel taller if we can assure ourselves others are small. And so, we keep racing. We race to be better, smarter, happier, healthier, and more successful.

We name our kids things like Apple and Atticus, because we wouldn't dare allow one other kid in their class to have the same name as them (no offense, Karen). Our kids are an extension of us, and we are special. We are different. We are so happy and fabulous. Except studies show our depression rates keep increasing. A study in *Psychological Medicine* found "the prevalence of depression increased from 6.6 percent to 7.3 percent between the years 2005 and 2015 with an even greater increase (8.7 percent to 12.7 percent) among those ages 12 to 17."[3]

I have unfollowed people I couldn't clap for. I remember that age-old adage, "If you don't have something nice to say, then don't say anything at all." The problem wasn't them; it was me. If someone is milking their work with missions for IG likes, that's on them. But the second I start a text thread scoffing, well, that's on me. Why are we competing with people that aren't competing with us?

I realized I hindered my own prayers by trying to block someone else's blessing. My refusal to just shut up and clap wasn't decreasing their success, but I do think it prevented mine. God isn't as worried about changing your circum-stances as he is about changing you. The best thing that ever happened to my faith was watching other people open packages I had ordered. I learned to smile from the bleachers, even though they wouldn't have noticed if I had walked out. I had to accept it wasn't my turn, but it didn't mean mine wasn't coming.

And so, I faked it. I started to say I was happy for them, even when I wasn't. I would literally say over and over, *Heather, it's not your turn.* And while something inside me started to die, something better was born. I started competing with myself, rather than with others, and in so many ways, started to truly live my life again.

The Focusing Illusion

Bestselling business author Simon Sinek gives a motivational speech called "Understand the Game," where he mentions a study asking people if they would rather have a $400,000 house on a street where all the other houses are $100,000, or a million-dollar house on a street where all the other houses are two million dollars. People chose the $400,000 house, even though it was lower in value than the million-dollar house, because they wanted to be better than their neighbors.[4] I truly believe comparison is the death of all our joy.

There is a theory that psychologists have described as the "focusing illusion." This theory suggests when we compare our lives to others, we often focus on small details and assume if these small details were different, we would be happier. For example, have you ever been having a fantastic day and then decide to scroll your feed? Suddenly, you come across another picture of Sydney. Sydney is perfect. Her hair is always perfect, her outfits are stellar, and her thighs don't touch (which honestly just feels unhealthy), but you are jealous, so . . . whatever. Sydney is on another vacation. Greece, hashtag Mykonos. Suddenly your perfect day is spoiled. You can barely afford the Olive Garden, let alone Mykonos.

What does Sydney do for a living anyway besides tag companies on Instagram?

Here you are trying to be faithful, trying to tithe 10 percent of your negative paycheck, and Sydney—who misspells *acropolis*—is in Greece for the fifth time this year.

Really, God?

A once perfectly happy life suddenly crumbles when you stop focusing on what you *do* have and start focusing on what you do *not* have. That is the focusing illusion.

Research into our comparison problem doesn't stop there.

Princeton conducted a study on college students, and the entire study consisted of two questions:[5]

► How happy are you?

► How many dates did you have last month?

The researchers found a weak correlation between the level of happiness of the college students and the number of dates they had been on. Then the researchers decided to try something. They flipped the order of the questions.

Now the survey read:

► How many dates did you have last month?

► How happy are you?

Suddenly, a strong correlation existed between how happy the college students were, and the number of dates they had been on. What happened? The only change was the sequence in which the students answered the questions. The second sequence forced them to change their focus. When a perfectly happy student focused on the number of dates they had—or didn't have—they no longer felt so happy.

Millennials are constantly bombarded with images of their successful peers. I can't tell you how many times I have finally felt on track with my life, and then I log onto social media and see my pal Andy Gerard on another trip to Kigali working hard in international development. Here I think I'm doing big things for the Lord, but I'm not in Kigali. I thought I was in the thick of ministry—until I compared myself to Andy.

In her book *Mythical Me,* Richella Parham explains that since social media is always curated content, when we compare ourselves, we are actually comparing ourselves to mythical players in a curated game.[6] The people we compare ourselves to are the best versions of those other people, not even who they really are. No wonder we struggle so much afterward.

Jodie Gummow, a writer for Salon.com, says social media plays a huge role in lowering self-esteem.[7] Apparently, two-thirds of people find it hard to sleep or relax after spending time on social networks. Of 298 users, 50 percent said social media had negative influences on not just their self-esteem, but their lives.

Psychologist Sherrie Campbell says, "When we look to social media, we end up comparing ourselves to what we see, which can lower our self-esteem. On social media, everyone's life looks perfect, but you're only seeing a snapshot of reality.

We can be whoever we want to be in social media, and if we take what we see literally then it's possible that we can feel like we are falling short in life."[8]

Determined to Be Better

Comparison can be a bad thing, but it can also be a good thing. I'm a millennial, so I would never tell you to shut down all your socials and go back to the Dark Ages. I love that I know you had sushi last night even though I know literally nothing else about you. I can't tell you how many times I am cackling and my husband says, "What's so funny?" and I'm, like, "My friend Dave said . . . ," and my husband is all, "Have you ever even met Dave?" and I'm like "Don't take this from us . . ."

Social media can be positive. I find it so annoying when people criticize the reality of our culture. Sure, warn me of the pitfalls, absolutely tell me to monitor my time usage, but if you tell me to go dark, I simply can't take you seriously. We should use social media as a tool, while warning people not to use it as a crutch. The same social media that makes me jealous when it's not my turn can also make me want to be better. I want to be more like Andy. I want to be more like my friend Vimbo, who has a nonprofit where she works tirelessly trying to build a school in Zimbabwe. I see the posts of my friends who are active in philanthropy and leadership, and it makes me strive to be better. I am so proud of my friend Jason Lemon (@JasonLemon), who writes articles for *Newsweek* covering our politicians and making people aware of issues overseas. I love watching my friend Scarlett (@ScarlettPosner) be vulnerable enough to talk about how motherhood carries an invisible load that disproportionately rests on the shoulders of women.

When I was in college, my dad used to say if I hung around three party-chasers, I'd be the fourth. But sit at the table with world changers, and you'll start to believe you can be one too. Social media doesn't just expose us to stupid TikTok videos. It also gives us Bernice King while scrolling Twitter from the couch. It put Steven Furtick on my YouTube and Alexandria Ocasio-Cortez making macaroni and cheese on her IG Live. Social media has helped me understand there are Black theological thought leaders and Generation Z members who are willing to take on big business, and LGBTQ Christians who continue to show grace to people who would rather pretend they did not exist. Social media hasn't just given me all of society's ills, it has also given me some surprisingly good answers. It has made people who I could never have met become real human beings. Sometimes all you see are @'s and # signs, but there are people out there making you a better lover of humanity from their cell phones, and I'll never go back.

You must understand how this works. Not only are we comparing ourselves to others, but others are comparing themselves to us. So, who are you? Not just offline, but online?

It's Not Your Turn

Maybe it's not your turn right now. Maybe you've been overlooked and underappreciated. Maybe you have ten bridesmaid dresses but no groom, or enough rejection letters for a bonfire.

Maybe you can't stomach another baby shower or typing the word "congratulations" one more time. What do you do when everyone else gets the move, the relationship, the success, and the accolades?

You show up anyway. At the end of the day, all we have in life is our integrity. Our followers won't get us to heaven, and our success and riches can't come into our caskets. However, a life lived intentionally can make ripples that continue long after you are gone. You can't control your circumstances, but you can control how you show up to them. Suddenly, I realized obscurity was a really safe space to grow, and I could stretch further if no one were watching. If I lived each day walking toward my destination, how would that change the way I went through each step? What if some seasons are temporary, and we can make ourselves better in the waiting room? What if you don't have to wait until it's your turn to live like your turn is coming?

I woke up one day and realized who I am when it's not my turn is more important than who I will be when it is. Anyone can stand on a stage for a crowded stadium. It takes conviction to get up when no one would have noticed if you walked out. I want you to get up. Not for them, but for you. Not to outdo Sydney, but to outdo yourself. Now, I'm not saying that if you manifest hard enough all your wildest dreams will come true. Some people never get the wedding. Some writers never get the book. And some singers never see a stage. There is no magic wand for life that can put bows around all our broken pieces. But what if we commit to the journey anyway? What would happen to who we are as people if we committed to do the work in the dark with no guarantee of light? What if we don't quit just because we're tired? What if we don't run to win, what if we run to learn? What if we do our best, not for raises, but to grow? Is it possible we can end up with something better than a happy ending someone else gave us? What if we finish our lives with a dignity we could only have given ourselves?

I want to be a real Christian who follows Jesus where he is headed, rather than tells Jesus to follow me where I am headed. That starts when it's not my turn for accolades.

In fact, there's no better place to start, than when *It's Not Your Turn.*

Promise one to memorize in a weary season:

"A good name is to be more desired than great wealth, favor is better than silver and gold." (Proverbs 22:1 NASB)

DISCUSSION QUESTIONS

► What are you waiting on?

► Talk about a time when you saw God bless the person next to you and you felt jealous.

► What experience can you lean into where God showed up for you? Could that same God still show up now?

It's Your Turn to Wait

Patience is not simply the ability to wait,
it's how we behave while we are waiting.

JOYCE MEYER

MY NANA HAD TEN CHILDREN. She was a single
mother, an African American, and lived in the inner
city of Boston. She had one dream: to get a college education. When you are a single female raising ten children in
the inner city as a minority, you don't get to chase dreams
like that. She had reality to deal with. It wasn't her turn to
chase after her passions. So she put her dreams into her
children. She made sure every single one of them attended
higher education. Please let that sink in. My Nana raised
ten children as a single mother in the inner city and every
single one of them went to college. Just because it's not

your turn doesn't mean you can't make sure it becomes someone else's.

When she was able to dream for herself again, her mind wandered back to her own aspirations. She had every reason in the book to settle. She had been a good mother, she raised successful children, and she stayed committed to her faith. She had done everything right, and no one would have thought she fell short if motherhood was her only accomplishment. And yet, in her seventies, my Nana attended Harvard University. Harvard. In her seventies.

In many ways, she is the reason I started studying stories. That one story of my Nana sitting in a Harvard classroom, or walking across the Harvard campus, made me believe that I, too, could dream crazy dreams, and God would bring me to my purpose if I could be patient. Patience does not come naturally to a millennial. It has been the sharpest sword I have ever held, and also has cut the deepest. I am now the wise age of thirty-three years old, and it has taken me thirty-three years to learn patience is the key to satisfaction, to wisdom, to happiness, and to faith. The only problem is . . . no one has it.

We can't wait. None of us. We don't watch television week to week, we binge entire seasons in forty-eight hours. We don't call people on the phone, we text we are almost there and then honk from the driveway. My generation is even killing cereal. No one wants to take the time to pour cereal into a bowl and deal with the potential mishaps that come with one hand on the steering wheel and the other balancing milk and a spoon. Hand me a protein bar and a yogurt shake and let me be efficient. We simply can't wait. Patience is a coffin. But what if patience is also necessary?

"You have all the talent," my mentor once said to me. "What you don't have, Heather, and what you cannot rush, is *experience*."

I have never forgotten these words. It reoriented my entire thought process. What if you aren't waiting on talent? What if right now you have what it takes to apply for that job? What if you already have what it takes to go on that date? What if the marriage, the career, and the promotion are all well within your reach? What if it isn't a matter of talent at all that God is stalling you on? What if you are waiting on experience? And what if experience can't be rushed?

What Is Patience?

Patience is our ability to accept or tolerate delay, trouble, or suffering without becoming angry or upset. Having patience does not mean you have to resign yourself to negative outcomes, it means you are taking power over your emotions in a delay. Patience is the ability to step back and regroup. It is the capacity to not see delay as a threat to where we are going. Worry and agitation accompany patience typically because we don't know if we will get the reward we believe we are owed. It's not just that we don't like delay (though that is part of it), it's that delay often increases our fear that this thing we have been waiting for will never show up for us. We will never get the relationship, we will never get the job, we will never get the house. Most of us could be patient if we believed these things were irrefutably still arriving.

I used to ask my friend Vimbo if she knew she would be married in five years would she be bitter with God right now? She would probably be able to finish grad school, travel, do the things that may be harder to do once she has a family. It is the fear God will never bring a husband that makes delay so difficult. What if your delay is not arbitrary? What if God isn't just giving my friend a wedding, but a husband and life partner that would be worth waiting for?

You Need to Imagine

I want you to do something with me. I want you to imagine. I want you to really think about whatever thing it is you feel your life is missing. I want you to picture the broken piece you have cried over, or that unseen area you feel overlooked on, and I want you to imagine what possible positive benefits may come from God delaying that thing. What if the marriage you want so deeply right now is delayed? What possible benefits could come from that? Is there anything you would do right now if you knew you had one year left to do it? What could be a benefit of you not getting the house this year? What if a better house and a more affordable house isn't on the market until next year? What if you would love the house, but hate those neighbors? What if there is more than you can see on this singular request? Imagine what possible outcomes may only be available to you through time and more experience.

In a study at UC Berkeley conducted by Adrianna Jenkins and Ming Hsu, it was discovered imagination may be the pathway needed to uncover patience. The study found when we imagine possible outcomes, it allows us to be more patient in a way that does not pull on our brain's need to use will-power.[1] The technique has been dubbed "framing effects" where you provide yourself with different scenarios as to how options are presented to you. Even when the reward is identical, the way we frame them can spark imagination, which in turn produces a willingness to be more patient. When Hsu told participants they could receive one hundred dollars tomorrow or $120 in thirty days, the researchers framed the option in what they called an "independent frame." You could have this or that, and the option was an independent choice.

Another way of giving the same reward was to provide a "sequence frame" to the option. Researchers told participants they could have one hundred dollars tomorrow and zero dollars in thirty days, or zero dollars tomorrow and $120 in thirty days. When we provide our brain with a sequence, we have a greater ability to be patient.

When we have an option to get up early or sleep in, the only thought process our brain typically engages in is what the immediate next consequence of not waking up will bring us. We don't think past that ten minutes of extra sleep. We use an independent frame. Get up now, or sleep in ten extra minutes. But what if we used a sequence to frame the option? We could sleep in ten extra minutes, or we could get up now which would give us ten minutes to make breakfast before leaving for work. We typically don't imagine, in that moment, what we could do with our ten extra minutes. We could meditate. We could eat. We could read. We could simply not feel the anxiety of rushing. The research indicates when we allow our brain a sequence of possibilities, it activates our ability to imagine what could be.

So, take a minute and imagine. Whatever it is right now you know *It's Not Your Turn* for, what could be a possible benefit of this delay? If Nana had gone to Harvard at thirty years old, that would have been amazing, but there is something special about a woman who won't quit even in her seventies. It says so much about who she was. She couldn't have possibly known her delay would be my inspiration. There is no way she could have known her story of waiting would give me the gift of patience two generations later. What if more than God is trying to give you success, he is trying to give you a story?

Delayed Gratification

The ability to wait may be one of the most beneficial character traits you can acquire. It could be why Christ is so patient with us while we develop it. James 1:2-4 (NKJV) reads, "My brethren, count it all joy when you fall into various trials, knowing that the testing of your faith produces patience. But let patience have *its* perfect work, that you may be perfect and complete, lacking nothing."

Why would James suggest trials, which produce the opportunity for patience, should be something we meet with joy? In fact, James takes it the next step further. He claims the work of delayed gratification, the power of patience, is what brings about the perfect work of Christ in us. The perfecting of your character happens in the waiting room. The way to take your spiritual life from toddler to maturity is through the power of patience.

Patience is holy work. We are wired for gratification. Freud called it the "pleasure principle," the idea being pleasure is instinctual. We seek pleasure and avoid pain. He claimed this need is the driving force controlling the most basic force of our nature. Freud opined, with children, there is only the pleasure principle. They are unable to do or think of anyone or anything aside from pleasing their own desires. This is what keeps babies alive. They cry to eat. They cry to sleep. They instinctively know their needs are important.[2]

It's good for you to be able to say out loud, "My desires are not bad, my desires are human." I sometimes worry that in Christianity we have come up with this idea that to constantly deny our desires is holy. That the more we suffer, the godlier we are. I want you to reject this teaching because while I think God can redeem suffering, I don't think he revels in it. One of the fruits of the Spirit is joy. Your happiness is part of your holiness. Your joy is godly. Of course, we can be plagued by sinful desires, but many desires in and of themselves are not sinful, and your pursuit of them is not unholy. God wired us with a pleasure principle. Churches where there is a lot of laughter are churches where God is near. We don't talk about that enough. Real, instinctual, authentic joy should be the bedrock of our churches. Sterile halls with hushed "amens" aren't what make us holy. Being living, breathing people who live out the gospel is. In Scripture, God designed his dwelling to move among the people. It is us who made it stationary. We shouldn't go to church; church should come with us, and joy should be what calls others near.

Studies do show that practicing the muscle of delayed gratification is one of the most effective traits of successful people. While pleasure is not bad, the maturity resulting from our developmental stage of learning how to delay it can be extremely beneficial to our career, our focus, and our relationships. What if learning to delay pleasure actually produced more pleasure later? For example, what if taking the time to go through a degree program provides you with a higher income? (I mean, it didn't for me, but what if it does for you?) Successful people learn to delay gratification in order to attain a higher goal.

A well-known study that chronicles this is called the "marshmallow test" arranged by researchers at Stanford University in the 1960s. In the study, children were put in a room and a single marshmallow was placed in front of them. One researcher would then provide a simple in-struction to the child: you can have one marshmallow now, or you can wait fifteen minutes and receive two marsh-mallows. The children who delayed gratification with the goal of experiencing deeper pleasure through the two marshmallows scored higher on standardized testing and were less likely to struggle with behavior issues.[3] Delaying pleasure may very well be the skill God is trying to give you because he knows how much it will enhance your well-being. God can change your circumstances in a moment. What God cannot change, because you must surrender it, is your will.

A 2007 study conducted at Fuller Theological Seminary found patient people were less likely to suffer from de-pression. Patient people were found to be more grateful and expressed they felt more connected to mankind.[4] That is quite the scientific finding. We see yet again how science confirms Scripture. James writes the perfecting of your patience will complete the perfect work of Christ in your soul. Science says patience will make you healthier, happier, and more connected to humanity.

A study by Sarah Schnitker revealed people who rated themselves as having the ability to be patient with people around them also reported being more satisfied with their lives than people who were impatient.[5] Schnitker also found patience was linked to hope. Patient people were more forgiving, they were more empathetic, and they were more equitable.[6]

"But let patience have *its* perfect work, that you may be perfect and complete, lacking nothing." Patience truly is the work Christ is trying to complete in all of us. God is not punishing you through this trial, he is perfecting you through it. He is not abandoning you in this wait, he is developing you in this wait. God is not angry and wanting you to suffer. God is loving and wanting to use this season to your benefit.

God can answer your prayer in a moment. But what God really needs are vessels he can keep pouring into. God knows how to play the long game. He knows what patience does, and in his goodness, he will allow you to strengthen a muscle that will carry you much further than this single destination. You think you just want a job; God says the career I will give you will exceed this one place of employment. You want a house; God says I am preparing you to be a shelter for thousands of other people. You want a marriage; God says the work I am completing in you will allow you to bring companionship to hundreds of people whom I can entrust you with.

I believe God is still in the business of answering our single prayers, but if you would be willing to wait, maybe he is trying to do so much more than that.

It's Your Turn to Experience Peace

This concept is particularly important to me because I feel like I have lived it. I know what it feels like to wait. I know what it feels like to believe so deeply that God is going to move in your life only to sit in a waiting room. I know the ache that comes with hope that literally hurts your shoulders to keep carrying. I know this space is hard. I know what it feels like to eat ramen for days, and have breakfast for dinner, not because you like breakfast food, but because pancake mix is cheap.

I know what it feels like to sit in an interview and feel like you really put your best foot forward only to wait for a call that never comes. I know what it feels like to tell people God is going to do something and then tell them you were wrong. I know what it feels like to watch a family member who you love more than anything get sick, and suddenly the love that is supposed to go on forever becomes terminal.

I don't ever want to dismiss how painful this life can be. I have cried so hard I couldn't breathe. I have wanted to run away from myself. I have felt such deep shame that I've fallen to my living room floor and honestly thought my heart would stop beating. I think a problem with Christian culture is the temptation to tell people to put ribbons on their grief because "God is so good." And yet it is God himself who says in Micah 1:8 (NKJV), "Therefore I will wail and howl, I will go stripped and naked; I will make a wailing like the jackals." What if even our grief is godly? What if it's okay to be devastated about our country? What if we are allowed to wail about what is happening in so many of our churches? What if we were to allow ourselves to table the upside, for just a moment, to strip ourselves naked and beat our chests in the streets? In fact, what if that is the piece we are missing?

Patience and waiting are easier when it comes to jobs and advancement, but I recognize some of us are waiting on far more intimate moments. My mother-in-law, Nicole, lost her nineteen-year-old son to cancer. As a mother myself, that kind of heartache is probably my greatest fear. And when I ask her, "How do you survive something like that?" she says, "In a way you can't understand until it's happening to you."

Nicole tells me she can't explain the way God surrounded her during her son Tyler's final days. She knew grief, she knew heartbreak, she knew desperate prayers and hopes of bartering with God, but somehow, in the pit of that lion's den, she also knew peace.

"As Tyler died," she says, "I felt the physical presence of God. I can't explain it. But it was unlike anything I've ever known." And I believe her. Not because she is my mother-in-law, but because the look in her eyes isn't just devastation at what she has lost, it's also hope in what she has found.

"Heaven is not a distant place for me," she says. "Heaven has a face, and it has a name. Heaven is Tyler."

The last thing Tyler said to my husband seemed like odd language for a nineteen-year-old boy to even say. "Brother," he said, reaching for Seth, "at all costs, you have to be there."

Tyler's death has had a lot of cost. Seth felt like it defined who he was for years. He was seventeen years old when Tyler died, and that type of trauma changes everything. There have been days my husband hated God and days he hid from him. But Tyler's words have always sat in the back of his mind: "Brother, at all costs, you have to be there."

Tyler knew what I think the rest of us have struggled to believe: whatever God doesn't make right in this life, he will make new in the next one. There is a bigger and better wait than whatever singular prayer we are waiting for right now. And whatever dreams I am waiting on, while important, become quite small when I look at Nicole. I can't fathom her wait.

Sometimes I'll ask my grandma to move to Denver with me. She always says no. "I lived with your grandpa in this house," she nods. My grandpa died over seven years ago. But she would rather sit where he once sat than go to a room he never stood in. She wants to wait for him there. When I picture my eighty-nine-year-old grandma sitting in an empty house, talking out loud, it brings me to tears. Heaven won't just be the climax of all our earthly dreams, it will be the filling of every empty room, and all these years we waited will fade into that reunion.

Philippians 4:7 says, "And the peace of God, which transcends all understanding, will guard your hearts and your minds in Christ Jesus." I think that's what Nicole tries to tell me about. A peace that physically guarded her, that transcended all understanding. I don't know how painful your waiting room is, but I know it can take everything you have just to try and stand up again. But you must, when you are ready, try to stand up again.

Because, at all costs, you have to be there.

> **Promise two to memorize in a weary season:**
>
> "Great peace have those who love your law, and nothing can make them stumble." (Psalm 119:165)

DISCUSSION QUESTIONS

➤ When have you waited on the Lord in the past, and it proved to be for the best?

➤ What is the fear that makes waiting on this particular prayer so scary?

➤ Do you trust God to keep his promise to you? Why or why not?

It's Your Turn to
Say It Out Loud

*Raise your words, not your voice. It is rain
that grows flowers, not thunder.*

RUMI

I WAS EXPELLED FROM SCHOOL IN EIGHTH GRADE. I'd
like to clarify that it was a small conservative Christian
school, so finding trouble wasn't difficult if you were willing
to look. I was not slinging dope at recess or sneaking alcohol
into the gymnasium. I was simply not wise enough to know
how powerful my own voice was. I didn't know back then
that words, even spoken by a twelve-year-old, could stir a
commotion. It's a lesson I've never forgotten. I said things
because I thought them. I willingly questioned authority if I

didn't agree with their logic, and I honestly didn't think this was a bad character trait (you can blame my parents).

I was never super disrespectful, but I think in the nineties, at least in my church experience, people didn't think kids had the right to ask questions or challenge ideas. This was probably the first sign I was a millennial, but no one had coined that term yet.

For example, once for a choir program, the music teacher selected a song where we were all to sing "cotton needs a' pickin' so bad." I read the lyrics and immediately complained. Why are you forcing a diverse student body including Black students to sing a song about picking cotton? Not on my watch. I created a petition and asked for a meeting. The teacher ended up pulling the song but was deeply offended I even brought it up. People didn't talk about race much back then, especially not in my small Christian school. Everyone busied themselves not seeing color.

Imagine a student poking holes in the system at her con-venience. I'm pretty sure this is why they expelled me. I was too vocal. I questioned the church, I questioned faculty, and I questioned the entire system if I thought it needed a second look. If I saw injustice (mainly just against me because I was twelve and kind of self-centered), I sounded the alarm. But in a small school, and being both female and Black, I don't think that made me seem like a "go-getter," which is how my dad kept reframing things. It just made me a problem.

One day, though I had never even been suspended, I found myself expelled. I don't think it was any one particular thing I'd said or done, they were simply tired of dealing with me altogether. When they expelled me—or as they referred to it, "strongly suggest I withdraw"—I remembered being incredibly hurt and confused. I was annoying, sure, but to be expelled? I'll have you know I never got in trouble again in public schools. In fact, at a parent-teacher conference, a teacher told my parents, "You can tell the church school to send us their bad kids any time." Billy *was* slinging dope at their recesses. They had real fish to fry. My complaining about the cost of a school field trip hardly made their most wanted list.

This is when I first realized the power of my own voice, and choosing to speak had consequences. Saying things people didn't want to hear could result in someone shutting you up. But it also launched me into what I believe is the most important concept I've ever learned—words have power, and our voices have influence. I genuinely learned this at the age of eleven, and it has been the gift that has only kept giving.

According to Joseph DeVito, the author of my favorite communication textbook, the definition of power is having the ability to change what someone thinks or does.[1] With this lens, I want you to pause and think about how powerful words are in everyday life. You have the power to open your mouth and influence what your spouse thinks or does. You have the power to compose one tweet and influence how your network thinks and acts. You have power to write down thoughts other people will read and be moved by.

Angry words have the power to shut down the brain's ability to provide logic and reasoning. When someone says something you disagree with, maybe that attacks your faith or political beliefs, two tiny organs sitting on top of your kidneys begin pumping adrenaline into your bloodstream. You never tell your adrenal glands to do this, they just do it. Suddenly in response, your brain will separate essential functions from nonessential ones. It sends blood to your big, high priority muscles. As large muscles like arms and legs receive more blood, the high-level reasoning areas of your brain receive less. Your body is prepared to save you from a literal physical attack, but your brain is no longer functioning with all its blood resources. You are under an immense amount of mental and physical pressure—all because someone knew the power of their words. Our human ability to communicate is one of the most powerful gifts to humanity, and here we often sit, unsure of whether we have anything worth saying.[2]

Words Change Your Brain

I read about an experiment in the book, *Words Can Change Your Brain* by Andrew Newberg and Mark Waldman. I am going to do it with you right now. But first, take three deep breaths, roll your neck three times each side, and when you are ready, keep reading.

No

NO

NO

NO

NO

NO

What happened when you read those two letters? For some of us, maybe we didn't feel anything at all shift, but just because you don't feel it doesn't mean something isn't happening. In fact, when I include this slide in communication seminars on how our messages impact our workplaces, and those two letters pop up on the screen, multiple people in the room tell me they can feel a small jolt. Their brains immediately react. They respond with something telling them alarm was present.

Newberg and Waldman will tell you that if you were connected to an fMRI, in less than a second, your brain would have responded to those two letters. Your amygdala, just now, produced dozens of stress-producing hormones and neurotransmitters in a matter of milliseconds: "These chemicals immediately disrupt the normal functioning of your brain, especially those that are involved with logic, reason, language processing and communication."[3]

If two letters can make your brain short-circuit, what do you think happens when we argue with our spouses? What do you think happens when you talk trash about Betty in the break room? What do you think happens when you call my husband after his sermon saying Jesus told you to share with him your notes on everything he did wrong? (Pastors' wives, where you at?) If two letters can disrupt our brain's ability to function, what do you think rabbit-holing over money, arguing with our kids and teenagers, undergoing spiritual abuse does to our brain's ability to process accurately? Two letters create dozens of stress-producing hormones. Our negative thoughts are probably one of the most powerful weapons the devil uses to immobilize us.

So, what are you thinking about? We have to stop lying to ourselves that our toxic self-communication isn't hurting anything—that it's okay for us to spiral down thought trails about how fat we are, disliked, stupid, or boring. We have to stop telling our brains we are never going to be as successful as so and so. It's not okay to watch videos of Meghan Markle on YouTube and wish we had her hair (sorry, that one was personal, but you know what I mean). One of the most important conversations you will have every single day is with yourself. If two letters thwart your brain's ability to think clearly, what do you think your self-hate speeches are doing to your productivity? This is why I started telling myself, *Heather, it's just not your turn.* My brain could register that it not being my turn right now didn't mean my turn wasn't going to come.

Rather than say, *Heather, you are never going to pay off all your student loans,* I started to say, *Heather, you aren't able to pay off your student loans yet.* That *yet* allowed me to acknowledge my reality without allowing it to be a definitive statement over who I am or my career.

Carol Dweck, the author of *Mindset: The New Psychology of Success,* says right on the opening page of her introduction what Newberg and Waldman essentially say in several chapters: "A simple belief about yourself ... guides a large part of your life. In fact, it permeates every part of your life. Much of what you think of as your personality actually grows out of this "mindset." Much of what may be preventing you from fulfilling your potential grows out of it."[4]

Our words have power. They have the power to keep us going or to get us to walk away. They have the power to make us believe or make us give up. But what if the characteristics you've been speaking about yourself isn't just who you are—the person who quits after difficulty and shrinks after rejection? What if that *isn't* your personality but rather the result of a belief you have been holding on to? And what if that one belief is wrong?

Your Words Change How People See You

Newberg and Waldman say that vocalizing negative thoughts creates even more stress hormones, not only in our own brains but in the brain of the person we are communicating with. "You'll both experience increased anxiety and irritability, and it will generate mutual distrust, thereby undermining the brain's ability to build empathy and cooperation."[5]

Our negative language creates distrust in our listeners. This is a concept I talk about a lot in communication theory courses, the idea being it is not only that our own words can shape us, but also that our own words shape how other people perceive us. It is not okay for us to just keep talking without thinking. We must take more ownership of our communication. The good news is we can train our brain to both control our communication with ourselves, and others.

Sometimes we think if we can plant a dirty seed about someone, it will taint how other people see that person. But what often happens is it taints how our listener perceives us. Wendy Patrick, a career trial lawyer, writes in an article for *Psychology Today* about a theory called "trait transference." Essentially, trait transference says what you say about other people becomes attributed back to you.[6] Apparently, even when we know the traits being transferred don't fit what we know about the person speaking, our brains still do it. If you describe someone as genuine and kind, the people you are talking to start to associate those traits back to you. But if you describe someone as sneaky or bitter, those same traits also get spun back to your web. Think about how this would work in your romantic relationship. If I tell my husband as he pours my coffee, "Wow, Seth, you are so loving. When you pour my coffee first every morning, I feel appreciated," Seth then thinks more positively of me because of the positive words I attribute to him. He transfers those kind words back to me. It's magic! So, when I yell, "Seth, you couldn't make my coffee first? You are so selfish!" he is also more likely to attribute the negative feeling of selfishness to me. Speak wisely. People are listening.

These spontaneous trait transfers are not logical. It is a mindless association that we perform subconsciously. Patrick notes in her article it isn't just that we transfer positive emotions; we, in fact, transfer specific traits. Meaning, it's not only that you say someone is a hard worker and people now have a positive impression of you, it is also the literal trait of being a hard worker their brains ascribe to you. The simple act of speaking kindly about the people you work with can make people think kind thoughts about you.

It doesn't stop there. Speaking kindly about others doesn't just improve an outsider's perception of you—speaking kindly to yourself can also improve how you see yourself. "The brain, it turns out," Newberg and Waldman write, "doesn't distinguish between fantasies and facts when it perceives a negative event. Instead, it assumes a real danger exists in the world."[7] So when we say, *I think my husband may be cheating on me,* we are telling our brains not what hypothetically may be a negative fantasy, but what is actually happening. Our brains will respond as if this is a real threat. It becomes as real to our brains as the ground we are standing on. The thoughts we feed our brains become the ground by which we operate. Negative thinking is a self-perpetuating system. We don't just have one negative thought, we create a dozen different thoughts we then ruminate over, which means more negative thinking, which means more stressful damaging hormones that could potentially physically change our brains.

The goal then becomes not to pretend things don't exist, or to be the creepy Christian who says everything is fine when the whole house is burning down around you, but to respond to our negative thoughts so our brains don't spiral. Rather than just saying, *I think my husband may be cheating on me,* you respond to that thought with, *and if he did, he would be sorry, because I am worthy of respect, faithfulness, and value.* If you don't respond, you keep spiraling. If you do respond, you change the focus from what may be to what you really are. So, *I am not doing good in my graduate course* gets a response of *and here are the action steps I will take for the next two weeks to increase my study time.* Don't just linger on the bad thoughts, respond to them, and interrupt them. Give your brain positive ground to stand on.

The Power of Scripture and Knowing Your Word

"Hope deferred makes the heart sick" (Proverb 13:12). Has hope ever made you sick? I know I have felt sick of hoping, sick of dreaming, sick of missed goals, and even tired of having faith. Faith can be beautiful when we are able to clutch our hands around it. I've certainly had days where only faith got me through. But when days hold hands with years, eventually even faith can become a burden. I have had moments where faith in God felt painful. I have been so sure of God's deliverance, only to sit on my porch like an eight-year-old waiting for their dad to take them to a baseball game and those truck headlights never come. My friend and spiritual accountability partner, Vimbo, has been praying for a husband for over five years. Two years ago, I told her to buy a tie and hang it in her closet. I thought maybe if she had a physical symbol of the promise she believed God would one day bring to her, it would help keep hope alive. But now she has had a tie hanging in her closet for two years. It just makes her angry.

Sometimes faith feels embarrassing. I feel Joseph on a personal level. "'Here comes that dreamer!' they said to each other" (Genesis 37:19). For years, Joseph held on to this notion that he would be some type of leader for his family. He was given dreams as a child, and he felt in his bones these dreams were orchestrated by the same God his father preached about. But after months passed, and then years, I am sure promises felt heavy. Depending on what your dreams are, it can even feel like friends and family are waiting for you to fail.

"Come now, let's kill him and throw him into one of these cisterns and say that a ferocious animal devoured him. Then we'll see what comes of his dreams" (Genesis 37:20). In a pit,

Joseph waited. If I didn't read my Bible, my own pit would bury me. If I didn't know that saints before me had dirt under their nails, I would be tempted to think something was wrong in my story. It would be easy to believe pits are where things eventually stop breathing.

I've sat in a pit. I called off my engagement two months before the wedding. I had a dress, I had invitations. What I didn't have was a man I was supposed to marry. Earlier that evening, I prayed God would intervene and end my engagement if it weren't his will for my life. Within twenty minutes of saying "amen," my fiancé called and broke up with me over something so minor I can hardly even place the reason. The God who created the universe answered a cry from a twin-sized bed. The God of brimstone and fire burned up all my plans for a future. I asked him to intervene, but I didn't realize he already had a lightning bolt in his pocket. I was in a pit, and I thought surely I would die there.

Two hours later, my phone rang. I assumed my fiancé was calling and decided I wouldn't turn back. If I had chosen someone God hadn't chosen for me, it was better to end an engagement than a marriage.

"Hello," I whispered.

"Heather?" he said.

It wasn't my fiancé. The voice on the other end of the line was Seth Day, my very first boyfriend I'd met in sixth grade. He had blue eyes and baggy pants, and while I wasn't sure what love was, I hoped it felt like buying Skittles after school with Seth. Seth moved away after eighth grade and I didn't see him again until I was a sophomore in college. I wrote my number on a piece of paper and said, "Call me sometime." He didn't. *Thanks, bro.*

Now here we were, two years later, on the very night I prayed for God to end my engagement if it wasn't his will, and Seth called me. I cried on the phone with him. He had stumbled into a hurricane. He offered to drive down to visit me the next evening. He came, I cried, and he listened. And then, at some point, he made me laugh. The next weekend he came down again, and the next and the next. Eventually, one day while sitting on a pier, he reached for my hand, and I haven't let it go since.

This story of how I met my husband is also how I know God is busy working out plans for my life. Had Seth called me instantly the day I gave him my number, I don't think we would have gotten married. I met my ex-fiancé a couple months after giving that number, and he was outgoing and charismatic unlike my shy and introverted husband. Sometimes, you have to experience what's bad for you before you know what's good for you.

There are two things about God and prayer I find to be helpful to remember, and the first is found in Daniel 9:23. It says, "As soon as you began to pray, a word went out, which I have come to tell you, for you are highly esteemed." Sometimes we pray, and immediately a command goes out, and God places the answer to our prayer in our lives. I would classify my experience with my husband in that category. For whatever reason, I prayed, and God answered in the exact same night. It's honestly unbelievable.

But there is a second example we find in the very next chapter, Daniel 10:12-13. The second prayer is not answered immediately, and it's interesting to read why. "Then he continued, 'Do not be afraid, Daniel. Since the first day that you set your mind to gain understanding and to humble yourself before your God, your words were heard, and I have come in response to them. But the prince of the Persian kingdom resisted me twenty-one days.'"

This verse is super important to understanding how prayer works, or, as it sometimes seems, isn't working. Bible commentator Charles Ellicott says of this exchange:

> Perhaps no single verse in the whole of the Scriptures speaks more clearly than this upon the invisible powers which rule and influence nations. If we were without a revelation, we should have thought it congruent that God Himself should direct all events in the world without using any intervening means. But revelation points out that as spiritual beings carry out God's purpose in the natural world (Exodus 12:23; 2 Samuel 24:16) and in the moral world (Luke 15:10), so also, they do in the political world. From this chapter we not only learn that Israel had a spiritual champion (Daniel 10:21) to protect her in her national life, and to watch over her interests, but also that the powers opposed to Israel had their princes, or saviors, which were antagonists of those which watched over Israel. The "princes" of the heathen powers are devils, according to 1 Corinthians 10:20.[8]

In Daniel chapter nine, we see God answer immediately, and in Daniel chapter 10, we see Daniel's answer is delayed due to the intervention of evil supernatural influences in the region at that time. What is important to note is that both times, Daniel is loved by God. A delay in this instance is not brought on by Daniel himself, but rather, is a reaction to the organization of demonic spirits in the supernatural realm. Sometimes, it is not your turn, and it is also not your fault. There is a real devil and a real army of evil constantly organizing to delay your promise. Pray through.

Either way, you can rest assured God plays chess, not checkers. He is not rushed, he isn't nervous, and he doesn't need to make a quick fix. He is looking at the long-term gain, while we can only see the short-term joy. But what comes quickly isn't necessarily what we need the most. Sometimes what we need is the very thing we can't see because it's a few months or years down the timeline. We can't see it, but God can, and he can't risk our ultimate joy with a quick fix resolution.

Before Joseph ever found himself in a pit, his dad had also waited. Joseph was the son of Jacob. Jacob spent twenty years in Paddan-aram. In Genesis 32, we see after twenty long, hard years wrestling with his wives, wrestling with his father-in-law, and wrestling with his birthright, Jacob the strategist, Jacob the heel-grabber, Jacob the symbol of our church, wrestled with God. Jacob spent twenty years sweating in a foreign desert. He spent twenty years in fear of his brother Esau. Jacob probably cried out to God many times in the dirt of the desert in prayers of going home. Jacob waited, and a person who waits is simply a person who hasn't given up.

It's funny how the desert often comes before the promised land. In Exodus 3:5, God meets Moses in the middle of a desert. He has fled Egypt, and for forty years he lives in the desert. When God calls Moses from the burning bush, he tells him to remove his shoes, for the place he is standing is holy ground. But Moses is in a desert. How can our hardest nights, how can our darkest trials, how can the deserts where it feels like our hope will be buried, be what God calls holy ground?

John the Baptist calls a nation to repentance in the desert. Right after Jesus is baptized, before he ever performs one miracle, he heads out into the harshness of the desert. Jonah has a desert, Paul has a desert, Mary has a desert, Martha has a desert, Job has a desert, the disciples have a desert. You show me a promise, and I'll show you a desert. If you have found yourself in the midst of the desert, you are in excellent company. God says to remove your shoes, for the place where you are standing is holy ground. What if who you are while it is not your turn is actually the foundation of who you will be when it is?

In the dark of the desert, Jacob wrestles with God, and we see the key to surviving the desert: never let go of Jesus. In Genesis 32:30 after refusing to let go of God in the desert, God blesses him. Jacob proclaims he will name the place Peniel, for he has seen God face to face. The desert is where we come face to face with God.

Today's Christian is being lost in the waiting room. We are programmed to think if God were real, he would move on our behalf instantly. He'd snap his fingers and fix it. He'd throw a lightning bolt at all my troubles and I will magically have faith and healing. If we want a deep relationship with God, we should be able to have one, and fast. And yet . . . Moses waited.

Moses doesn't even begin his mission until he's eighty. He is a baby when God protects his little raft through the Nile, and yet it is eighty years later that Moses finally hears from God. According to Exodus 12:40-41, the people of Israel have already been afflicted over four hundred years. Surely, if God took the time to send an angel to protect baby Moses in the Nile, a holy storm was brewing. God was clearly moving, so why the wait? Could it be that it would take Moses forty years to process everything he had been taught in the Egyptian palace? In Moses' case, the weight of his leadership is so pivotal for the trajectory of Israel that God has Moses spend forty years in the wilderness preparing him and the people. What if your own wait isn't wasted? What if what God is leading you to will require a wilderness experience? What if the splendor of your own Egyptian training has to be stripped by the stillness of a desert? Remember that Christianity is truly an upside-down kingdom to American culture. It is possible that this waiting period isn't arbitrary.

Joseph is sold into slavery for at least thirteen years before he is placed in charge of Egypt. What if it took thirteen years for Joseph to become the man who would forgive his brothers? Joseph was sold into slavery as possibly a spoiled teenager, but when his brothers find him again, he is a deeply wise and forgiving man. He is the man that would move all of Jacob's house into Goshen, which is the same move ensuring Moses is positioned in the great city of Egypt. What if

Joseph wasn't just waiting for his own deliverance? What if he is the key to everyone else's? What if God letting him into the palace too soon would have given us an entirely different story? What if it wasn't Joseph's turn, not because his turn wasn't coming, but because who he was when it wasn't his turn created the person he would be when it was.

Here is why you must develop patience: perhaps the strangest story of waiting is that Jesus Christ himself waits thirty years to preach one sermon. Even Jesus waits. Jesus. Mr. Perfect, Mr. No Sin, Mr. God Incarnate . . . waits. There must be something pretty special about time. Time must edify us. Our entire culture is set up for instant gratification, yet Christ tells his own disciples that while for them, any time would do, his time "has not yet come" (John 7:6 NKJV). In fact, in Luke 2:52 it says, "And Jesus grew in wisdom and stature, and in favor with God and man." If Scripture says that Jesus "grew in wisdom" that would also mean that there was a time Jesus did not know as much as he would come to know. When Jesus decided to become a man, it meant he was also committing to the process of development. Jesus grew. Isn't it quite egotistical then, for us to not submit to do the same? I trust God with my growth, and the amount of time it takes to prepare me for where I am going.

If Jesus Christ himself believed there is something powerful and transformative in time and patience, why would we be exempt? Of course, the reality is some of us may die in the desert. Israel waited for four hundred years. Many people in that time never saw the prophecy fulfilled. Which is why at some point I started asking myself, what if there is a deeper purpose for the promised land then the one I am imagining? And what if the greatest part of my story isn't how it ends, but how it continues. What if this life is actually a small thing when compared to our eternity? John the Baptist died in prison, but his eternity is secure.

The Power of God's Voice

There are over 3,500 promises for us in Scripture. Remember what Newberg and Waldman said: our brains can't distinguish the difference between fact and fantasy. Our words are as real as the ground we are standing on. What if God is not just asking you to read your Bible because it will make you holy? What if he has been trying to arm you with over 3,500 promises that could change your mindset? What if God knows words have regenerative power? What if the God who created the brain knows what you'll need to defend it? What if he wants you to be able to disarm the negative attacks of Satan with the promises of God?

But our brains cannot recall what we haven't read. We cannot respond with promises we do not know. I don't know what attacks you are daily living under. I don't know what hope you buried in your desert, but I do know Isaiah 41:10 says, "Do not fear, for I am with you; do not be dismayed, for I am your God. I will strengthen you and help you; I will uphold you with my righteous right hand."

I know Isaiah 40:29 says, "He gives strength to the weary and increases the power of the weak." I know a few chapters later, Isaiah 54:17 says, "No weapon forged against you will prevail, and you will refute every tongue that accuses you.

This is the heritage of the servants of the Lord, and this is their vindication from me."

I believe Moses of all people, a man who knew waiting, a man who knew hardship, a man who knew what it felt like to feel buried in your own desert, I believe him when he wrote in Deuteronomy 31:8: "The Lord Himself goes before you and will be with you; he will never leave you nor forsake you." What if the promise of God is the presence of God? And what if the promised land is wherever God is, even if it's a desert?

I encourage you to read your Word. I have been weighted by negative thoughts about who I am and what I can do only to hear the still small whisper of the Holy Spirit answer with the truth of who I am and what God can do through me. I have literally said, "I am not strong enough, smart enough, funny enough, popular enough, eloquent enough to ever be anyone special." And I heard a whisper say, "I am. It's not about what you are, it's about what I am. I am big enough, strong enough, smart enough, eloquent enough to walk you through this. I AM." And suddenly I understood why when Moses asked God, "What should I call you?" God said, "I AM" (Exodus 3:14). It has never been about what you are, it is about what God is.

What Are You Saying Out Loud?

I told Vimbo, the one who has had a tie hanging in her closet for two years for the husband she hasn't met, something I want to share with you. I started telling her out loud, "I know you can't believe this right now, so I am going to believe it for you." When she can't pray it, I do. And I say it out loud because every time I speak it, I believe it a little more. When she can't see it, I do. And when she can't believe it, I believe it for her. I pray every morning for God to lead her to her person. Most nights before I go to sleep, when I pray for my own husband and kids, I include hers. I say it out loud. I speak it because I know how much she wants it, and I ask God to hear our collective voices and answer a prayer she may be too tired to pray.

Six months ago, my single friend who has never even had a boyfriend—I have been praying for her for over five years—started dating someone. I don't know if he's the person the tie belongs to, but I do know he tells her every day how beautiful she is, and how lucky he is to have found her. He is the same guy who has sent her random texts since the day he first met her four years ago saying things like, "Whenever I meet someone who knows you, I get excited that I am able to bring you up, I am so proud to be your friend." They have known each other for four years, went on two dates back then, and nothing really happened with it. But here they are, four years later, after bonding over phone calls and pandemic quarantine, turning their friendship into something romantic. Today, when she sent me screenshots of her texts with him, I asked, "Do you think this is the guy?"

"I don't know. We talk about marriage. I hope he is the guy," she responded. "But I do know that I've never had a man make me feel this beautiful before, so even if he's not the guy, I am so grateful for how he is changing how I see myself."

I recognize I am a hopeless romantic. I know I tend to swing to a glass that is half full. I know I'm the friend who tells you the upside, who thinks of what could go right rather than rattling off everything wrong. But I also think everyone needs just one person in their corner. One friend who can see it when you no longer see it. One person who can pray when you don't feel like you can anymore. One person who is willing to say it out loud when you are afraid to hear how it sounds. I don't know what you can't see past right now, but I want you to say it out loud. I want you, for just one second, to let yourself hope again. Let me be your positive friend.

Let me believe it for you.

Promise three to memorize in a weary season:

"Thy word is a lamp unto my feet, and a light unto my path." (Psalm 119:105 KJV)

DISCUSSION QUESTIONS

► What negative thought traps do you often loop through?

► Do you have a belief about yourself that you have said is your personality but may just be a deeply held belief? Has this been positive or destructive?

► Explain in your own words why it is important we know our Scripture and engage with it daily? How could it protect our thoughts?

4

It's Your Turn to See

*The relation between what we see and what we know
is never settled. Each evening we see the sun set. We
know that the earth is turning away from it. Yet the
knowledge, the explanation, never quite fits the sight.*

JOHN BERGER

I SPOKE WITH MY MENTOR on the phone this week, and
he asked me, "What do you see?"

I told him right now I am too tired to see anything.

"I'm exhausted," I said. He didn't respond.

"Do you know how hard it is to give your life to ministry
and constantly have people tell you it is either not enough
or wrong?" I heard crickets.

"My family and I took a job halfway across the country because I believed God told me to, and nearly every moment

since I have come, I have worried I made the wrong choice," I said.

"Hmm," he finally made a sound.

"José," I continued, trying to get some sympathy, "I had a home in Michigan. The housing market in Denver is nothing like Michigan's, and I am afraid God won't make a way for me to ever afford a house out here. I am living in a campus duplex, 900 square feet, with three kids and two dogs who are all now quarantined together due to coronavirus, and I have no idea how I am supposed to keep writing lectures and grading assignments and preaching via Zoom seminars."

"I have heard a lot about what you feel," he said. "But I asked, what do you see?"

My husband answered, "I see we are living in campus housing when we could be paying a mortgage we can't afford at a time when the housing market is probably on the verge of collapse."

"I see," I said, taking a deep breath, "the car in our driveway was literally given to us by a woman we didn't even know."

A few months before this, my husband had been having issues with his truck. We had just moved to Denver and we were extremely low on funds because neither of us had been paid for over two months by our last jobs, and our new payroll hadn't started yet. Our house in Michigan still hadn't sold, and we were counting on that money to get us through the transition of our move.

On top of all this, my husband needed a new car. We didn't have the money to fix his truck. We cried most nights feeling as though we had accidentally gone backward in life. Aren't your thirties supposed to be peak years? I had a hot tub in Michigan. In Denver, I couldn't afford Jimmy John's.

One day, Seth saw a woman named Volinda standing in the fellowship hall after church. He had just started his first pastoring job, and when he saw her, he said he heard a still small voice whisper, *Go talk to her.* He thought she belonged to our church, but she didn't. Everyone looked new to us. She was just traveling through. Her brother had died, so she and her elderly mother had come to clean out his home. They googled churches in the area and ended up at ours because it was off the highway. She said it was very difficult—because it was just them—to move all the boxes and sort through her brother's belongings. My husband felt this was why God told him to talk to her, so he could help. So, he did. He went over for the next three days and cleaned out the house of a man he never met, for a woman who didn't even go to our church.

One day, Volinda asked, "Do you like Starbucks?" Seth had brought one over each morning he came. "Yeah," he responded, embarrassed.

"Here," she said pressing a gift card in his hands, "my brother just put $270 on this gift card before he died."

When I tell you, this particular month my husband and I were low on cash, I mean it. I was too embarrassed to ask my parents for help because they didn't want us to move in the first place. I had spent our last few thousand dollars on Christian education for all three of our children, which also was triple the price of what it cost us in Michigan. We had no savings left and no paycheck coming for at least two more weeks. That $270 gift card fed my entire family for fourteen days. We ate Starbucks breakfast sandwiches and protein boxes for breakfast, lunch, and dinner. That small gesture of kindness from Volinda, who didn't even know us, was enough. But she did more.

Her brother had a Jeep she intended to sell for $3,500. Seth told me that if we had the money, he would buy the Jeep because it was in great condition and his truck's air conditioner wasn't working anymore. This was August, and he was tired of driving places in suits that were full of sweat by the time he got there. Not to mention, the truck had a slew of other issues. The day before Volinda went to leave, she asked Seth if he wanted to buy the Jeep, though she had another buyer already lined up. I think she noticed the sounds coming from his truck.

"I don't have the money to put down on it right now," he said, embarrassed.

"How about you write me the checks, and I won't cash them until you say I can," she answered.

He couldn't believe it. He wrote her $3,500 worth of checks, and she gave him the title. About two weeks later, we got a letter in the mail from our angel Volinda, who we thought we had been sent to help, but who had clearly been sent to help us.

I feel like God has told me to give you the Jeep, she wrote. *Thank you for all you did to help my mother and me.* Tucked into the letter were all the checks Seth had given her, voided.

I had never seen God do something like this for me before. I had seen him in dozens of small ways. I had read about him doing crazy things like this in Mark Batterson's book *The Circle Maker* and I remember reading his stories and thinking, *Wow. Is a God like that even real?* and now here I was, complaining about everything I felt without acknowledging everything I could see. We have lived in Denver eighteen months, and my husband, who is serving his first eighteen months in full-time ministry in a church with less than two hundred members, during a pandemic, has had ten baptisms. So, while I feel homesick, I have to *see* baptisms.

What do you see right now? For a moment, pause everything you feel. Our feelings are wonderful ways to lead us, but they should not be lord over our lives. Some days, I feel God rush over me like a mighty wind. Other days, I feel nothing. If anything, I feel forgotten. But what I feel about God, and what I know about God, are often two different things. I have had to learn God will provide, even when it doesn't feel like it. I have seen him bless my husband's ministry, even when we didn't feel blessed. I have seen him provide me a safe space for my family to live and a car for us to drive when we couldn't afford it otherwise.

Another time, a few months later, my university's former vice president Kyle Usrey called me into his office. It's not every day a vice president of a university calls a brand-new faculty member to chat. In fact, in all my time in academia, I never had this happen before. Honestly, I was nervous he was going to fire me or something.

"I am going out of town for a while," he said.

"Oh," I responded, unsure of why he shared this with me.

"I want you to keep this." He handed me his campus dining card. "You'll never be able to spend it all. But I want you to try. Take Seth, take your kids. Have dinners in the cafeteria anytime you want."

I kept my composure, but inside I wanted to fall onto the floor. There was no way he knew we were struggling financially. I only talked to him one other time when I first took the job. I couldn't believe what God did. Here I prayed for God to bless me financially, and God said, "I'm going to give you a house on campus, I am going to put a car in your driveway, and I am going to give you all the cafeteria food you can eat." It wasn't the glamorous blessing I had been praying for. It wasn't cash in my wallet or a raise for my paycheck. But it was a humble blessing I will never forget. Sometimes God takes us around our problems, but other times he takes us through them. *Oh, God, open our eyes, that we may see.*

Eventually, we did catch up financially. Our paychecks stabilized, our house sold, and while we still spend most of our money on Christian education and student loans, we were also able to save for a down payment on a house, which honestly, we only bought to the testament again of God's mercy. My mentor wanted me to stop talking about what I felt, so I could talk about what I could see. When I think about what I see, the answer is the hand of God over my life.

I want you to take a moment and think about what you see. What do you see right now? What spaces make no sense if not for the hand of God? For just a moment, I want you to step back from everything you feel so your brain can make room for what you see. The thing about feelings is they can override your brain's ability to think logically. Let's take fear, for instance. If you are afraid to lose your job, afraid your spouse is cheating, or afraid you can't trust your coworkers, these fears produce a chain reaction in your brain. And while these feelings may be legitimate warnings you need to know who to trust, without logical reasoning to guide them, they can also make you paranoid without justified cause.

Fear produces hormones in your amygdala that send messages to your prefrontal or sensory cortex. The cortex alerts your hippocampus which starts comparing this current threat causing the fear response to past threats. Since the hippocampus is where we store memories, it is able to determine—and quickly—if the current danger is life threatening. This is why in the famous testimony of Christine Blasey Ford, who accused Supreme Court Justice Brett Kavanaugh of sexual assault at his Senate Judiciary Committee hearing, she talked about the hippocampus when asked how she remembered something so long ago.

Huffington Post writer Anna Almendrala recounts Blasey Ford's testimony in her article when she writes:

> "The same way that I'm sure that I'm talking to you right now, just basic memory functions," Blasey told [Senator Dianne] Feinstein in response. "And also, just the level of norepinephrine and epinephrine in the brain that sort of, as you know, encodes—that neurotransmitter encodes memories into the hippocampus, and so the trauma-related experience then is kind of locked there whereas other details kind of drift."[1]

And then Blasey Ford says the line that news stations would play over and over for weeks. It's the line the entire article was titled after. Almendrala quotes her, writing:

"Indelible in the hippocampus is the laughter, the up-roarious laughter between the two," she said, referring to Kavanaugh and Mark Judge, the other person Blasey alleges was in the room when the assault took place, "And their having fun at my expense."

Our brain is always storing memories, and memories are often created when we experience events deeply tied to emotions. These emotions help us create memories that are stored in our hippocampus, just like Blasey Ford talked about in her famous quote. Right now, if I say, "Think about your first kiss," your brain can immediately bring back an either positive or negative memory you haven't thought about in years. If I ask you to think about your first heartbreak, no matter how many years ago, it all comes flooding back as if it were yesterday. Your brain is always comparing current experiences to the catalog it has stored of past experiences.

You can see why this can be problematic. Just because a past boyfriend or girlfriend cheated on you doesn't mean this one will. Just because you lost your job once doesn't mean you will lose one now. But our brains don't know that, and they can respond with past fear responses that are not necessarily warranted in this current situation—and sometimes I wonder if we don't respond to God based on past failed relationship information our brain has stored. If your dad left, that doesn't mean God will leave. In fact, he makes a promise to us in Deuteronomy 31:6: "Be strong and coura-geous. Do not be afraid or terrified because of them, for the Lord your God goes with you; he will never leave you nor forsake you."

God promises he will never leave us. And yet, most of us struggle to believe this. We see God as someone who will distance himself if we make a mistake, or worse, someone who is waiting for us to make a mistake so he can say, "I told you so." And it's not necessarily because we are bad people or faithless. It's because we have past experiences of being left, which make it difficult for us to envision a God who stays. I personally think the God who made our brains is merciful toward us in how they work. God knows if you have had a pattern of abandonment, faith may be harder for you than someone who had perfect parents and a perfect childhood. God doesn't condemn you for a lack of faith; I believe God is, instead, actively trying to provide your brain with a new experience. But with God, we can only experience what we surrender.

If we don't constantly ask ourselves why we feel what we feel, and never ask, "Does what I feel match with what I see?" we may be putting the actions of man onto the character of God. Since man is so often untrustworthy, many of us never fully allow ourselves to trust in God. Not necessarily because God is scary, but because trust is scary. And God is always asking us to trust him. Meanwhile, our brains shout: "Trust nothing, you have been here before!" Lord have mercy on our past experiences and how they shape us.

While the human brain is incredible, it also can be highly problematic if we aren't aware of how the science works. According to a University of Minnesota research article, "Once the fear pathways are ramped up, the brain short-circuits more rational processing paths and reacts immediately to signals from the amygdala. When in this overactive state, the brain perceives events as negative and remembers them that way."[2] Because emotional responses heighten your brain's ability to store a memory, it may now store pieces of negative experiences that will resurface every time you encounter something that reminds you of that past, previously stored experience.

Let's just say your spouse is late from work. This in and of itself is no big deal. But if you have had a previous relationship with someone who did not understand your worth and cheated on you, suddenly, your spouse's (who may very well be trustworthy) tardiness will bring forward a flood of past negative emotions. Triggers are real, and they impact us sometimes at the moments we least expect it. A dad who never kept his promises, a mom who called you names, a past romantic partner who cheated, a pastor who sexually assaulted you—all of these negative experiences often leave us with various triggers we transfer into totally new experiences. And this isn't because we are broken or defective, it's because this is how our brains store memories.

When I dated my husband, he asked me to not tell him I was done, or the relationship was over, unless I meant it. His brother had died. His dad left. "Leaving" wasn't something you bluffed about. It was a trigger for him. We have to be mindful of people's pasts and how we may activate those triggers in their present. And we should know what our own triggers are so they don't take over our future.

Your brain is responding based on your past stored experiences. It is trying to protect you, and that is great, unless the current situation is nothing like the last one, and your brain is still responding in the same way. Every day we make choices, we create wiring in our brains we then tell people are simply "habits." But these habits are not just distant things we happen to do regularly; they can become very deeply held ruts that are natural patterns our brain communicates and works within. This is why it isn't easy to just stop doing something you don't like. Your brain likes to keep its pattern and will react in distress when you tell it to break it, even if that pattern is bad for you.

According to the University of Minnesota article, one simple emotion like fear can weaken our immune system. When your body experiences fear, it shuts down what it believes to be nonessential functions. Fear is a great stimulus for threats, but it is horrible if it becomes a state of being. Fear can literally impair your long-term memory, and if it damages your hippocampus, it can leave people in a perpetual state of anxiety—every time they bring up past memories to check their current fears, their damaged brains reinforce this sense of imminent threat.

Fear can make us impulsive and often influences the decisions we make in negative ways. And of course, people who have suffered horrible traumas that left them in deep states of fear for a long time are likely to develop mental health disorders like anxiety, depression, and PTSD.

The more I understood about how my world-sick brain functioned, the more Scripture felt like a world that could set me free. Now when I read verses like Ephesians 4:23-24 (Worldwide English) which says, "Have a new mind and heart. Be a new person. That new person has been made like God. He does what is right and holy because he knows the truth," I am like, *Wow. I want that!* And I honestly believe God has been in the process of renewing my mind for the last ten years.

Spiritual practices like meditation and prayer are not just imaginary ointments—they are neurologically transformative. Even eight weeks after beginning meditation, your brain can show significant changes. Practicing meditation for thirty minutes or longer and for many years, shows incredible changes to your brain. Spiritual practices like meditation can literally enhance the brain's thickness and neuroplasticity. As human beings age, things begin to wither: muscles get smaller, sometimes our height shrinks, and the cerebral cortex thins. Just like working out adds mass to your muscles, meditation adds thickness to your brain.[3]

Students of meditation are able to repair these thinning cerebral cortexes. People who meditate develop brains so well trained, they are able to generate compassion in situations that would be difficult for a nonmeditator to show compassion in. One study detailed how meditators responded to people in a waiting room. Researchers discovered that 50 percent of people who meditated gave up their seat for someone in a waiting room, where only 15 percent of people in the control group (nonmeditators) did the same. Meditation, even when not focused on things like compassion, made people more likely to make compassionate choices.[4]

"Have a new mind," Ephesians 4:23-24 (Worldwide English) says, "and heart. Be a new person. That new person has been made like God. He does what is right and holy because he knows the truth."

I think Christians think having a new mind is a meta-phorical expression, but neuroscientists like Newberg and Waldman say this is biological. It is hard for me to understand why any Christian would want to keep this to themselves or not seek out biblical meditative practices. The gospel has the power to truly give people new minds. So, I want to ask you, what do you see?

In 2 Kings 6:9 (NKJV), Elisha the prophet is leaking useful information about a potential battle to the king of Israel. Every time the king of Syria tries to ambush Israel, Elisha gets a message from God and they avoid the attack. "Beware that you do not pass this place, for the Syrians are coming down there," he says. The king of Syria is frustrated that none of his attempts to overthrow Israel are working. Then one of his servants who apparently has heard of Elisha says to him, "Elisha, the prophet who *is* in Israel, tells the king of Israel the words that you speak in your bedroom" (2 Kings 6:12 NKJV). So the king of Syria sends horses and chariots and a great army in the middle of the night to go and capture Elisha.

First, isn't it funny how the king, thwarted several times by Elisha when he tries to attack Israel, somehow thinks he will be able to one-up Elisha in person? Why in the world, if his servant just told him Elisha even knows what he talks about in his bedroom, would he think he would be able to somehow get the surprise attack on Elisha now? Because our brains constantly compare our current experiences with our past ones. This can be bad for people who have been riddled with negative experiences, but it can also create an unrealistic belief about oneself for people who have been surrounded by power their whole lives. Most people are not good at having a realistic view of themselves.

Research suggests most people believe they are better than average.[5] We think other people can develop disease, but we can't. Other people can get divorced, but we won't. Our memories can also be biased toward us in ways that serve us best. You may remember the one time you won more than the dozens of times you didn't, and that stronger memory of winning can give you a false sense of security, making you believe you will win again. We do this even with games of chance. Some people will never speak up front because they have one bad memory of a traumatic event where they felt they humiliated themselves, so every opportunity to speak brings back the negative response. And yet other people without traumatic experiences are more likely to think of themselves as being better than they really are.

And here we have the king of Syria, who has already lost to the prophet Elisha, unable to see himself as someone who loses anything. He sends horses and chariots and soldiers to go and get the prophet. And when Elisha's servant sees the Syrians surrounding the city, he cries out to his master, "Alas, my master! What shall we do?" (2 Kings 6:15 NKJV).

I am sure he responds to Elisha in fear, and you already know what fear does to our brains. It triggers past negative experiences, it shuts down our unessential bodily functions, it causes our brain to lose the ability to think clearly. Fear can make us literally freeze. And in these moments where we feel fear, our brains are even more likely to go on autopilot. They are more likely to take the quickest brain circuit route. For some of us, depending on our habits or patterns, that may be to just run. Run, without even knowing where we are going. That is why some of us will even run from good things because even good things make us feel fear if bad things are all we know, and the possibility of something good creates a fear of what we can't control. Some of us will run from good relationship development, run from a job opportunity, run from people who are trying to help us because fear activates whatever our past pattern is. I had a student who lived in a halfway shelter for several years. She came to my office one day in a panic. We worked toward this for years, and now she was moving into government housing. Fear was her reaction to moving out of a shelter. I was stunned.

She said, "Struggle is all I know. Who will I be without it?"

I didn't understand it then, but now I do. For those of us who mainly pray *Help!* it becomes who we are. We are in a constant state of reliance on God to meet even basic needs. And what if one day that changes? What if one day we don't need as much help? Who are we then? Sometimes our struggles can be a part of us so much that it becomes hard to imagine a life that isn't defined by that area of brokenness. And I think we should talk about that more. Stepping into even seemingly good things can be scary if you've never known goodness. And it should make us incredibly sad that for some of us, it's not just that we know struggle, it's that struggle has become who we are.

My husband was a runner. I can remember going to the beach with his mom when we were dating, and she said,

"Seth struggles with commitment." She wasn't lying. Every time things got hard or scary, he ran. One time in college, he didn't know how to distance himself from a girl he was seeing, so he literally got in the car and moved from Michigan to Tennessee. He was just gone. Ran. Sometimes patterns displayed for us as children become pathways we start to subconsciously make in our own brains. Seth's dad didn't stay and raise him, he ran. And now here was Seth, years later, wanting to be nothing like his father, and yet following in his same patterns. And remember that patterns, once we create them, can be hard cycles to break.

Seth and I have been married for over ten years. And outside of one breakup that lasted three days in the first nine months of our relationship, he has never left me. And I don't think it is because I am so amazing. Runners run, even from amazing opportunities if they scare them enough. He has stayed because God gave him a new heart and mind. He started undoing all his negative patterns. None of this happened over night, but it did happen. And I've seen my husband now, once an eighteen-year-old boy who would experience fear and run, somehow grow into the kind of man who feels fear, and yet stays. God can give you a new mind and heart. But we also have to create new memories for our brains to respond to. At some point, if a runner wants to stop running, they must give their brain a memory where they stayed.

Elisha doesn't run when he feels fear. Not necessarily because Elisha is more holy than his servant. It could simply be Elisha has a brain wired with experiences where God provided. Elisha is the same man mentored by the prophet Elijah, who literally shows up with Jesus at the transfiguration. Elisha has seen the hand of God before.

In 2 Kings chapter three, Elisha sees God bring water in a desert. In 2 Kings chapter four, Elisha sees God fill empty jars with oil. In 2 Kings 4:8, Elisha sees God make a barren woman pregnant and in 2 Kings 4:32, he sees God bring a dead boy back to life. It's not just that Elisha was holy, it's that Elisha had all of these miraculous experiences he saw, and so when the servant sees an army about to overtake them, Elisha sees the same God who has provided for him all along.

> "Don't be afraid," the prophet answered. "Those who are with us are more than those who are with them." And Elisha prayed, "Open his eyes, Lord, so that he may see." Then the Lord opened the servant's eyes, and he looked and saw the hills full of horses and chariots of fire all around Elisha. (2 Kings 6:16-17)

I am going to ask you to do something. I want you to start asking God for a new mind. I want you to start asking that he provide you with new experiences that will forever change what you see. I pray this prayer every day: *Lord, give me eyes, that I may see.* I don't want to see with my broken eyes. My eyes always see what's missing. My eyes are quick to see a nine hundred-square-foot duplex, and not the Jeep sitting in my driveway. My eyes will often see the future that is so unknown, and not the steps of my past God has so delicately guided. But I want the eyes of Elisha. I want to put away my mind of fear and walk in the new heart and mind the book of Ephesians says is awaiting me. I want to live a Deuteronomy 31:6 kind of life that allows me to "be strong and courageous." I want you to claim the promise "Do not be afraid or terrified because of them, for the Lord your God goes with you; he will never leave you nor forsake you."

My mentor asked me, "Heather, what do you see?" Now, I am asking you.

What do you see? My hope is that even when *It's Not Your Turn*, you can see the hand of God.

Promise four to memorize in a weary season:

"Have a new mind and heart. Be a new person. That new person has been made like God. He does what is right and holy because he knows the truth." (Ephesians 4:23-24)

DISCUSSION QUESTIONS

► What do you feel?

► What do you see?

► What past experiences do you think inform how you *feel* right now, and also how you *see* right now?

It's Your Turn to Think Small

You need to let the little things that would ordinarily bore you suddenly thrill you.

ANDY WARHOL

I WAS HAVING LUNCH with my friend Tacyana. "Your God is too big," she said, casually, before taking another bite of her salad.

"I thought God was supposed to be big?" I said.

"Heather," she said, putting down her fork so she could really lean into me. "My God is small. The way you describe God makes me think you believe he is only concerned with the big things in your life. My God is concerned with the small things. My God isn't waiting for me to climb some mountain. He is with me just as much in this lunch, as he will be when I walk downstairs and go back to my office. God is helping you find your keys. He has time to make sure you take a nap. God is wanting to be a part of every single conversation you think isn't noteworthy. He knows it's the thousands of small moments that lead us to a single big one. My God is small."

Tacyana was a volunteer chaplain at Princeton and a pastor at a local church nearby. She had the most baptisms in the history of the church she served. In her four years pastoring, she brought in 58 percent of the membership. I could go on about women in ministry, but I'll just let her ministry speak for itself. There were these awesome pictures of her with people in the baptismal tank while she was pregnant. Now I knew why her ministry was so successful. Tacyana thought small.

Tacyana went on to tell me that one day she decided to believe she was anointed, and if she was, then her presence somewhere was never arbitrary. It was a divine assignment. If someone bumped into her walking down the sidewalk, she didn't see it as coincidental. She stopped rushing past people and started seeing them as intentional moments God ordained just for her. "It's exhausting to live this way," she said. "But I've seen God, and I won't go back."

This conversation Tacyana and I had in October of 2018 changed my life. I started living my life as if each tiny, insignificant moment were significant to God. From that lunch, I walked across campus to teach a class. I had about seven students, and they all looked tired. Faced with what appeared to be a disengaged class, I may have been tempted to dial back. Just stick to my notes and not take anything up a notch. Why come out strong to an audience that clearly just wants to survive the next fifty minutes?

But all I could hear was Tacyana's voice in my ear: *My God is small.* I prayed in my head, *Lord, this is your moment. These are your students. Use me.* And then I fought. I fought for that lecture like it was the only thing I had to do that day. I went big in a small space, and I watched all of them wake up. We had an excellent class period.

When it was over, one student asked if she could come to my office. She had never asked to come to my office before.

"I felt God speak to me in class today," she said. "I felt him give guidance on what he wants me to do next with my schooling journey. I just wanted you to know he used you." I was stunned. I was exhausted from pouring my heart out in an otherwise mundane Tuesday lecture, one of many other lectures I would give that week.

But I've seen God, I could hear Tacyana saying, *and I won't go back.*

Prayer of Surrender

I want you to think small. I want you to start living life as if each moment is not arbitrary but calculated by the greatest intellect this universe could even fathom. I want you to believe God has a plan and a purpose for your life. And God can't have a plan for your life if he doesn't have a plan for your days. I believe this is why Jesus always woke up early in the morning to pray. He asked God what the plan was for that day. I try to follow this model for myself. I cannot leave my house without time with God because God wants to direct my steps, even on mundane Tuesdays.

But before we go any further, I want you to do something I think is important in order for God to fully work in you like he would like to. I want you to surrender.

I went to a talk given by my friend and evangelist Jonathan Leonardo. He walked me through the importance of asking God, in prayer, what we need to surrender to him. Jonathan said every person is a vessel God would like to use. And because every person is a vessel God would like to use, every person should be treated as someone who has the potential to have all of heaven divinely able to work through them. Every person has value because every person is made in the image of God. This is why we should care that right now thousands of immigrants are in detention centers. That right now there are thousands of immigrant children who are saying they were sexually assaulted while in our custody.[1] We have to care about this, because every single one of us is a vessel with the potential to have all of heaven divinely able to work through us! The Lord does not see as we see. We see citizen and noncitizen, whereas God sees a global church and a child of God. I also want to remind you there is not a single thing you could do to make yourself unworthy of God's pursuit. His strength will be made perfect in your weakness.

Take a moment and ask God what you need to surrender to him and write it down below. Maybe it's fear. Maybe it's ego. Maybe it's insecurity. Ask God what thing is preventing him from working through you in the way he would most like to move through you. Ask him—and then be silent and listen. Write down what you hear.

Today I surrender _____.

Is Bigger Better?

In thinking about goals and the idea that so much of what we are able to believe is possible is actually based on past experiences, you start to realize when we think we have little faith, it's really that we haven't had enough experiences where our faith was built. This is why even our faith is something God helps us build, more than us naturally exhibiting it. You have to have an anchor point allowing your brain to say, *I have seen this before.* Belief is a muscle in the brain we need to exercise in order to see it grow. If you set a large goal for yourself but have no anchor point in your brain to support that large goal, you may not ever have the belief to really feel like this is possible. And that's not necessarily your fault. Sometimes bigger isn't better. This isn't about breaking your goal into smaller chunks. It's about literally just setting small goals. The idea is to give your brain practice seeing you achieve something. The more your brain exercises this muscle, the heavier the goals you will be able to lift. One small step at a time.

I want you to start thinking small. I want you to build your belief muscle in your brain. God is going to take you to big places but not until you can appreciate the small ones. I had a student in my public speaking class share with us that she had performed in numerous shows on Broadway. "I was a spoon in *Beauty and the Beast*," she said. Later she would go on to get bigger roles, and eventually for one show, she got the role of Belle. "But the thing about show business is you don't deserve to be Belle if you weren't willing to be a spoon."

How many of us are willing to be spoons? Rather than say, "I am going to become the CEO of a major corporation," maybe just start setting the goal that you will treat today like it matters in your small office cubicle. Maybe today, think small: "I am going to be a spoon."

A *Time* magazine article discussed how taking a few minutes every day to simply savor a moment rather than rush through it actually made people happier.[2] Slowing down while you shower, not turning on the television while you eat your dinner, putting your phone down while you have a conversation with someone—these tiny choices that seem almost insignificant, aren't. People who practice savoring the small stuff—only two to three minutes of reflection—show significant increases in their happiness. Tiny tweaks lead to big changes.

In *Psychology Today*, Christopher Peterson writes that wealthy people may have a harder time slowing down than others. Participants were asked to eat a chocolate bar, and—unbeknownst to them—researchers were measuring how long it took to eat the chocolate.[3] The longer the time spent eating equated to a person's willingness to savor a moment. If the person was primed with a photo of money, they spent less time eating the chocolate bar and were less likely to want to savor a small moment (researchers controlled for gender because women were more likely to savor the chocolate in general, which, honestly, is so on brand I can't even comment).

What are you priming your brain with? Are you focused on success and finances? This may seem innocent, but it could also prevent you from savoring small moments which other research has already linked to your happiness. Your quest for the next better thing is directly correlated to how much you enjoy whatever you are in right now. Bigger is not always better. Be a spoon.

Think Small

In a *Business Insider* article, the founder of Rubicore, Nana Dooreck, wrote six small things the best leaders do differently, and what would you know: thinking small is a leadership trait. Great leaders think about the small stuff, even their own language. They are more likely to replace the words "I" and "me" with "we" and "us," and people notice.[4] This is a small change you can set as a goal for yourself immediately. How can you be more conscious about sharing the glory and acknowledging people's contributions? In bestselling author Jim Collins's book *Good to Great,* he says good leaders and bad leaders either use mirrors or windows. If something goes well, a bad leader looks in the mirror and attributes it all to himself. If something goes bad, they look out the window and cast blame. But for good leaders, it flips. When things are going well, they look out their windows and give credit to everyone who worked hard, and when things go bad, they look in the mirror and take responsibility.[5]

In researcher Liz Wiseman's book *Multipliers: How the Best Leaders Make Everyone Smarter*, she says there are talent magnets and talent blockers. She writes, "One VP had a favorite saying, quoted often and written on her door: 'Ignore me as needed to get your job done.' This simple mantra signaled an important trust in the judgment and capability of others."[6] Great leaders are able to call out and shine light on the talent of others. They are not threatened by it. They do not give empty praise; they give specific praise that will allow you to believe what is being said of you. Wiseman says in *Multipliers* that some leaders make you realize how smart *they* are, while multipliers make you realize how smart *you* are. Leaders who inspire others literally multiply the work and effort of everyone around them.

Small leaders think about how they treat others and according to Dooreck, they also remember small details. Dooreck says an easy way for anyone to become more charming is simply to remember and recognize little things about the people around you. When you remember someone's grandkid's name, or anniversary, you are telling them you see them. You are acknowledging their value and importance in your world. Start asking yourself while it is not your turn: What ways can I better serve the people around me in this season?

One day, I was having a really bad day. I realized I couldn't control what was happening in my life, but I could make a choice to do something that would make someone else's day go better. I literally thought in my mind while driving to work, *Who are three people I can make happy today*? And then I went after them guns blazing, trying to assault them with sunshine. I bought my coworker Lynn a coffee. I stopped at the grocery store on my way to work and got flowers and a card for Deserene who singlehandedly held up our graduate department. I pulled aside my student Anna and told her I saw so much in her. I was having a bad day, but I realized I could still influence the people around me for the better, and of course, it influenced me. I can still remember driving home and feeling peace. I was grateful I didn't have to have a perfect life to make a difference. Honestly, I didn't even have to be happy to choose to bring happiness. And neither do you.

This brings me to another point on Dooreck's list: great leaders pay attention to the least important people. For some of us, this comes naturally because we know what it feels like to feel unimportant. We root for the underdog because we view ourselves that way. It is important for us to make sure everyone feels included, and I think even if that doesn't come naturally to you, you can train yourself to start responding this way. If God is going to make it your turn soon, don't you think he would want you to notice people who often go unnoticed before he gives you a larger reach?

I used to have an administrative assistant named Chloe who I don't think would ever say it, but I could tell didn't like me that much. Since it was difficult for me to accept she didn't like me (let me love you, Chloe!), I set out on a two-year journey to kill her with kindness. Every time I walked in the door, I told her good morning. I asked her the next day how things went that I remembered she stressed about the day before. I brought her coffees regularly and a card on her birthday. Within months, she became one of my closest friends in the office. And the closer we became, the better the office environment became for everyone. I sometimes sat out and ate my lunch while talking to her at her desk, and before I knew it, two or three other colleagues were joining us. My small decision to change my relationship with one person spilled into the entire office. By the end of the year, I had never felt closer to my coworkers. We accomplished incredible tasks and supported one another in a way you can only do when people genuinely care about each other. It was beautiful to be a part of, and it all happened because of one small decision to not blow past one person.

Another attribute Dooreck lists is something I can really speak to as a communication professor. Good leaders don't just have a single communication style. They can adapt to different people and speak in ways others are best able to receive. This is what we call "competent communication." Competent communicators can change their message to fit their audience. When I was in graduate school, I called my district manager because I experienced some conflict with a subordinate. "She doesn't listen," I said. "I tell her what needs to be done, and she literally just doesn't do it. She's a bad worker."

My district manager, Lyndsey, said to me, "This is your job. Your job as a head manager is to figure out how to get people beneath you to work well. I think you think her performance is her problem, but I am telling you it's yours. You have to figure out how to motivate different people to work for you, and what works with one person, may not work with another."

I was in a master's program for communication, but this manager in retail taught me something my professors hadn't.

She taught me that figuring out how to change my communication style could change my relationships dramatically. I had been talking to this employee the way I would like someone to talk to me: directly. But she wasn't me. She was sensitive and an indirect communicator. The way I spoke to her gave her anxiety, and she didn't know how to get her job done when I was around. I had to adjust my communication style. I started pointing out things I did like about her. I started affirming her more than criticizing what was going wrong. Our brains need compliments at a 4:1 ratio to criticism. This simply neutralizes out the single criticism. If you want someone to feel affirmed, you really need to hit a 5:1 ratio.

I started doing this with my employee, and I watched her blossom. Apparently, she wasn't a bad worker after all. I was a bad manager. It's a small thing, really, to change how you would typically respond in an effort to better connect to someone else. It's a small thing that changes how people perceive you.

Intercultural Communication

One of my favorite courses to teach is intercultural communication. If giving a presentation at work, American culture wants to hear about future benefits, whereas in India or China, you may gain more audience approval by establishing past achievement. Americans tend to be low context communicators. We talk about the facts of the message. Whereas Japan is a high context communication culture. You would really pay attention to the body language, time of day, and relationship in order to fully understand the context of the message. The more globalized our country becomes, the more interculturally aware we must be. Often, someone's words reflect them far more than they reflect their feelings toward you. Direct communicators go through life directing. They tell more than they ask: "Come in, have a seat." They direct you through language. Indirect communicators ask more than they tell: "Would you like to come in? Would you like to take a seat?" An indirect communicator may misunderstand the intentions of a direct communicator and take it personally. But their communication pattern is about them, not you. And a director may feel uneasy or distrusting of an indirect communicator. They may feel like you are trying to manipulate them rather than just giving it to them straight. But their communication is about them, not you.

I have experienced intercultural communication on a personal level. My dad is Black, my mom is White, my sister's husband is Asian, and my best friends in Denver are Latinx. There are just so many differences simply based on our cultures. I grew up going to Black churches and White churches. If my husband is preaching, I like to sit near him and talk back. If he says a good point, I shout "Yes!" or "Preach!" My friends here in Denver, Josue and Carla Vivanco, were raised in a very conservative Latin church community. No one even clapped during a church service, let alone shouted. When my husband used to come over to my family's house for lunch when we were dating, it was very loud. I would argue with my dad across the table, my sister would shout that both of us were wrong. My brother's laugh would shake the whole house. And my White mother had adapted to this environment and was the first to cut everyone off when she had something to say. My husband was overwhelmed.

"Why is everyone yelling?" he asked.

"Yelling?" I was so confused. "We are passionate!" My dad never felt like I was yelling at him at that lunch (trust me, the lunch would have been over if he had), we were just passionately discussing an issue together. Passion was allowed. Elevated voices showed engagement, not disrespect. This is what family lunch always looked like for me.

When I went to my husband's house for lunch, I suddenly understood. You could hear the fork touch the plate. No one spoke until the person speaking was clearly finished. People smiled and were friendly, but no shouting to communicate engagement. This is what family lunch looked like for my husband. You can imagine in our early days of relationship how this felt to navigate. He's loud now though, so don't worry about him. He adapted. Microevolution or something.

Growing Up Biracial

Growing up biracial has given me a unique vantage point on communication, on racism, on grace. It has taught me grace is not a small thing. My grandpa walked me down the aisle. He didn't get to do that for my mom. He didn't get to because he didn't go because my dad was Black. I've always instinctively known people are capable of better. No person is a singular action or response. Bad behavior can find redemption and repentance, and my interracial family has taught me that. My grandpa was the dad who didn't go to his own daughter's wedding before he was the loving man I always knew.

As racial tensions have heated up in the United States, I became angry. I am angry, and tired, and anxious. I am tired of White supremacy, and tired of Christians who don't find it to be a deal breaker. But then I remembered my grandpa. I remembered people are capable of better. I remembered grace is for people who don't deserve it.

My grandpa was one of the best men I knew. He used to carry a copy of my first book everywhere and show it to strangers at coffee shops. "My granddaughter wrote this," he would say, and they would have to pretend to care. He was the first in line to vote for Obama. He was so excited! I'm glad my dad didn't write off my grandparents, and I'm glad my grandparents repented. My dad didn't have a dad growing up, and my grandpa became a dad to him. I'm grateful for my biracial identity because there are days I feel disgusted at people's actions, words, or apathy, and God will remind me that before my grandpa was the man I knew, he was the dad who didn't go to his own daughter's wedding.

My grandpa died seven years ago and, in his memory, I've committed to patience, to love, and to doing what my own father decided to do when faced with injustice. He walked with them. He walked down the dusty, long, broken road of racial reconciliation. Because grace is for people who don't deserve it.

It's Not Your Turn

It may not be your turn right now to plan the big moves. You may be waiting on that one big investor, that one big phone call, or that humongous opportunity. I know what it feels like to exist in small spaces and not appreciate them. I wasted so many years of my life hoping for bigger and better, or faster. I had to stop waiting for my turn to come and start acting like it was my turn right now, even if no one else noticed or cared I was there. This space matters. This is your training ground. God can change your circumstances tomorrow. If that call comes in five minutes, what good is it if you can't be trusted with the responsibility? I want you to let go of the ladder, and just savor the ledge. The choices you make right here in this small space determine who you are in the next one. It's not your turn today. But it will be. Which is why you have to start thinking small.

> ### Promise five to memorize in a weary season:
> "One who is faithful in a very little is also faithful in much, and one who is dishonest in a very little is also dishonest in much." (Luke 16:10 ESV)

DISCUSSION QUESTIONS

► What is one thing can you do differently to start thinking small?

► What did God ask you to surrender? Were you surprised (or not) by what you heard?

► Can you pinpoint a small moment in your life that set you up for a big moment? Share.

6

It's Your Turn
to Set the Goal

*It's what you practice in private that
you will be rewarded for in public.*

ANTHONY ROBBINS

O NE TIME, I ASKED MY STUDENTS to share their first
vivid memory. One said a baseball game her dad took
her to when she was little. Another talked about a family
vacation to Disney. But one said, "Unclasping the metal clip
on my seatbelt as I tried to fall asleep in the back of my
mom's car as she went into the bar." Before I am anything, I
am a teacher. Nothing has impacted my life, or changed me,
like meeting hundreds of students every semester and seeing
how different they all are from each other.

What my time in a classroom has taught me is just because we have all ended up at a similar place doesn't mean we had the same start. I have seen that a room full of people can all come and enroll at the same school, at the same time, for the same class, and yet be worlds apart. I say this because I know before we talk about goals, we may all have ended up at a similar place, but that doesn't mean we have had the same start. For example, experiencing a trauma can often lead to real changes in someone's psychology. I had a student who had her house blown down by a tornado while she clung to her mom and brother in a basement cellar. The entire house was reduced to nothing but rubble. Miraculously, they all survived due to a doorframe in the basement left standing which sheltered them. I had her as a student several years later, and she still struggled to complete assignments if she was in her home. Once she became stressed, even over writing a paper, her cortisol started pumping and all she could think was, *I'm not safe here.*

Our ability to pursue meaningful goals is pivotal to how we view our lives, but the reality is there are outside factors affecting our motivation and goal-setting abilities. One single, traumatic event changed how my student processed and handled stress when in her home for many, many years. Sexual abuse, natural disasters, physical trauma, experiencing violence—all of these can change how motivated we find ourselves and our ability to pursue goals even if we have the motivation to make them.

For example, in order for goal reaching to be effective, research has found parallels between being able to pursue goals and how high your self-esteem and determination is. Setting a goal is about helping yourself. One quality of those people who achieve goals is they are good at impulse control. In context, we know people who have encountered trauma may struggle with impulse control. They may experience fear and helplessness as a result of the traumatic experience. We may not even realize the difficulty it may cause in the mental psyche of someone who feels helpless, because quite literally, at one point, they were. A woman who is sexually assaulted may have a harder time showing initiative in many other areas of her life because helplessness becomes a pivotal emotional response to stress. Or a veteran unable to save a friend or comrade in war may have the emotional response of failure become a pivotal response to other non-war-related situations.[1]

While goals can be a fantastic and awesome tool to improve your life, it is also important to acknowledge we haven't all had the same start. Give yourself some grace for the experiences that may hinder your motivation and goals. Everyone's journey is so complex and unique. One size never fits all. Right now, before we go on, think about what experiences may potentially hinder your ability to stay motivated? Or your impulse to feel helpless? Or your deep feelings of failure? Not everything we carry is ours to carry. Sometimes, they have been placed in our bags at the hands of people and experiences we have gone through. Maybe you aren't where you want to be right now. What potential setbacks in your life could be contributing to that? Maybe you aren't "lazy"; maybe you are healing.

What You Practice, God Will Make Perfect

Often people will say God doesn't call the qualified, he qualifies the called. I believe that, but I also know from my own experience, God will allow you to practice what he is about to make perfect. When people ask me, "How do I know where I am going?" I always tell them to look at where they have been. I think God often ties traces of your future into the fabric of your past. For me, the first time I ever preached I could hardly even breathe. I spoke at a Bible camp in front of about twenty high schoolers. I felt like I had an out-of-body experience. My anxiety was so high and my adrenaline so fierce, I wasn't even able to eat until it was over. I have no idea now what I even said. I had notes up front with me, but I couldn't read them. The pages actually blurred. Everything happened so fast.

I am so grateful God gave me that sandbox. He gave me a safe space to mess up and let my mind go blank. He gave me a contained area to learn how to process nerves up front and how to feel comfortable hearing my voice control a room. What if the first time I ever preached was in front of two thousand people? I am so grateful in public speaking, God allowed me to go from one sandbox to another. It probably took me four years of speaking regularly before I ever spoke in front of two thousand people. By the time I did, I knew I could. I knew I was good at this. I was able to be myself because I had grown into that arena, little by little, sandbox by sandbox. And you know what? Before I ever preached, I taught. God had given me classrooms to explore public speaking. I had the most honest audience—college students—who had no problem telling me if they didn't want to be there. I see very clearly now how teaching prepared me for speaking. Jeremiah 12:5 says, "If you have raced with men on foot and they have worn you out, how can you compete with horses? If you stumble in safe country, how will you manage in the thickets by the Jordan?" Little by little. Sandbox by sandbox.

In high school, I wrote for the school newspaper. I was an editor. Before that, in elementary school, I would write plays and then ask my teachers if our class could perform them. Once or twice a year, the entire class would perform one of the plays I had written. In college, I again became an editor for the school paper. I learned how to conduct real interviews and stick to deadlines. By the time I wrote my first "real" book at twenty-two, I had been writing for a while, just not for mass audiences. The only reason I believed I could write a book was because I had learned how to engage in thoughtful consistent writing, and the little steps of my past were actually the building blocks of my future. We don't thank the God of "little by little" nearly enough. We often complain because we think he is holding out on us. But what we practice, God is making perfect. If we can't run with men on foot, how can we run with horses?

David doesn't just wake up and become king, he first has to tend sheep. This is all in preparation for Goliath. David doesn't just take on Goliath; Scripture says he first takes on the lion and the bear. Then, after he defeats Goliath, he has to work in the army of Saul. And then after he gets crowned king of Judah, he waits seven years before becoming king over all of Israel. When Moses leads the Israelites in the desert, God doesn't just give them the Promised Land, he says something fascinating in Exodus 23.

> I will not drive them out before you in a single year; otherwise the land would become desolate and wild animals would multiply against you. Little by little I will drive them out ahead of you, until you become fruitful and possess the land. (Exodus 23:29-30 Berean Study Bible)

We don't just serve the God of big moves; we serve the God of "little by little." Start paying close attention to what common thread unifies your experiences. God is in the fabric.

If right now you feel God is taking too long, he is probably developing you. He is shaping you little by little because he knows you are valuable. The worst thing that could happen to any of us is we would become an instant success or an overnight sensation. No human should go from nothing to everything in moments. Your brain takes time to adjust to changes. Your thought processes will take time to adjust to thinking and creating with increased pressure. Whatever goal you are wanting to meet, if God is involved, expect a process of little by little. Is what you are in right now actually a sandbox? A nice, contained area where you can make mistakes and get dirty without much at risk? First we run with men, then we run with horses.

One of the most difficult things that can happen to a mother is her child be born too early. Premature births reduce survival rates. Whatever you haven't been birthed into yet, maybe it's because you are still being developed delicately and quietly in the womb of experience?

Anders Ericsson, a cognitive psychologist, studies experts and how they achieved their world-class skills. Malcolm Gladwell made Ericsson's research famous when he cited it in his book *Outliers*. Essentially, Ericsson found it took 10,000 hours of deliberate practice to become an expert. A stipulation Ericsson makes about his research is it's not just that experts practice longer, but they practice deliberately.[2] They set goals for themselves and practice until they reach them. For example, running every day doesn't necessarily make you a better runner if you don't set goals to identify what *better* even means. Expert runners train to meet goals. They run to cut seconds. They have a metric and measurement for success.

It is not enough for you to just continue on knowing you want to be better at something. What does better look like? What do you really want to accomplish with your life? What macro goals are you hoping to reach, and what micro goals could you set this week, month, and year to get you there? Often people will say they just want to be successful without giving their brains any tangible goals to measure what success means for them. What does success look like for you, and what is the metric by which you'll define success?

In the book *Daily Rituals: How Artists Work*, Mason Currey describes the habits of creatives and scientists. He says the common theme of successful people is that they are deliberate with their routines, and they practice them daily.[3] Playing piano once a week is one thing; playing piano for twenty minutes every day is another. We saw a similar theme in our meditation research, that daily rituals would have far greater rewards on the brain. Even being specific with where you practice can help your brain create automatic habits. During graduate school, I had a small corner desk on the top floor of the library. I wrote every paper at that desk. If I had a paper due in twenty-four hours, I knew it would take me twice as long to try to write that paper at home. I could have less time, and do more, from that small desk in the library. Why? Because my brain associated that space with work. It became automatic. I wasn't at the library to scroll Twitter. I was there to complete assignments, and my brain knew it.

Setting SMART Goals

Goals are the object of a person's ambition or effort, an aim or desired result. The very first step to getting where you are going is consciously thinking about what goal you have in mind to get there. Only 8 percent of people who set New Year's resolutions will achieve them.[4] Marilyn Price-Mitchell writes in *Psychology Today* that goal setting is linked to traits like confidence and motivation, and also just the small act of writing down a goal will make you 33 percent more likely to achieve it.[5]

Many educators talk about the term "SMART" goals, an acronym to help students think clearly about what they want to achieve. I want you to set a SMART goal if you haven't already. Paul J. Meyer explains the SMART process in his book *Attitude Is Everything*.[6] Some of the explanations he uses are as follows:

"S" stands for *specific*. You want to set a specific goal that you will put effort into achieving. After I finished my PhD, I decided I needed a new goal because my only goal for the last five years had simply been to finish my PhD. I prayed and asked God, "What's next?" I decided God's hand was in the thread of my life. I started to look backward. Here I had just finished a PhD, what door did I want a PhD to open for me? I had been writing inspirational Christian living books for several years. I had a pretty popular blog in my small community, but I had never written something research oriented or academic. I decided that was my goal: to be published with a national publication that was also academically focused. With SMART criteria, you set something specific.

The "M" stands for *measurable*. How will you know when you have made progress toward the goal? How much? How many? What is the metric? Having something defined that can be measured will keep you moving toward the goal. Even rejection, in some cases, can still be progress. If the measure is you will write five hundred words every week—even if they are five hundred bad words—your commitment to keep writing is still measured as a success. In my case, I decided I would query three publications a week with pitches for the next month. Rejection came every single month, but I kept pitching, and so I kept moving toward my goal. My goal wasn't just to be published; it was to keep pitching so I could eventually be published. When I changed how I saw my metric, I succeeded even if I failed.

The "A" stands for *achievable*. Is your goal to win the lottery? Or is your goal to pay off your credit card debt? The point of a goal is to set something you can put effort toward achieving. How is this goal achievable, and at the same time, difficult enough that you would have to put in effort? I knew it was achievable for me to write academic content because

I had a PhD. I tried pitching to the *New York Times, Psychology Today,* the *Washington Post*, and other outlets.

The "R" stands for *relevant*. Is this goal relevant to your success or life? Is this effort worthwhile? Is this the right time to try and work toward something like this? For me, the goal was extremely relevant because I had just finished a PhD program; therefore, I set a goal I saw as the next steps as to how I could use it. People often set a goal like writing a book, but they really don't have a story to tell, and possibly not even a history of writing. Our goals should be relevant to who we are and where we are going.

The "T" stands for *time-bound*. This is also what helps you really narrow down your specific timeline. What target dates do you have for when you would like to see certain things accomplished? I would literally set a deadline for my action plan. For me, I needed to pitch at least three places every month. That helped me to stay focused on the goal each month, even if I was busy with other projects. It also helped me to keep moving forward, which truly is one of the main things to help us reach our goals at all. I want you to really think about what you can put in your "T" section below. The first thing I always do when making a new goal is ask:

- ► What can I do today?

- ► What can I do tomorrow?

- ► What can I do in a month?

- ► What can I accomplish within six months from now?

Answering these timeline questions are critical. They give your brain a roadmap so it will automatically and subconsciously continue collecting information to help you reach these goals once you've written them down. After a few months of zero publishers banging down my door asking me to write for them (I know, weird, right?), I stumbled across my breakthrough. I scrolled my Twitter notifications, and I realized David Kinnaman, president of the Barna Group, tweeted me. I knew exactly who the Barna Group was because I grew up reading their research. I was probably the only sixteen-year-old in my high school who used the Yahoo search engine to read about church statistics.

I had tweeted the day before about an article I read, and someone asked me where the article was from. I couldn't remember, but I assumed it was Barna data, because I read Barna a lot. The next morning at 6 a.m., I opened my Twitter and saw the president of Barna, who had no idea who I was, nor was he following me, somehow had seen the tweet and responded.

"Heather, this is not our study," he said, "but let me know where it came from. It's interesting data." I totally fangirled. I fangirled over a researcher, which I recognize makes me a whole new level of nerd.

I clicked on his Twitter profile, and low and behold, the last post on his newsfeed was a call looking for data storytellers. (A) My entire dissertation was on storytelling, and (B) I had just spent three months querying national publications looking to use my academic research skills to make a larger contribution. Oddly enough, I had never thought of Barna. I wrote David back saying I would find the citation, and "Also, I see you are looking for data storytellers. I will apply." He said he would love for me to send in my resume. About six weeks later, I got an email saying I could contribute to them as projects came along. I was thrilled and have written several pieces for them.

Here is the thing. Had I not written down my goal of writing academic pieces, I would never have seen that post on David's newsfeed and even thought it applied to me. The only reason I recognized it was because my brain subconsciously swept in information to help me attain the goal I had written down several months prior. I am going to keep it real with you—I had stopped sending out queries. After two months of rejection, I was kind of over it. When I saw that call for writers, I froze. I couldn't believe it.

I had been writing with Barna for about nine months before they published one of my articles. They were releasing the latest research and my piece finally hit the social media circuit. My friend, a journalist for *Newsweek*, texted me: "Hey, I saw your Barna piece, why don't you submit to my editor? We need more Christian contributors who aren't staunch conservatives." I queried his editor that day, and my first article ran in *Newsweek* within forty-eight hours. Had I not written for Barna, I wouldn't have been able to start writing for *Newsweek*. Had I not written down my goal, I don't think I would have had any of these opportunities.

Setting goals matters. My husband has a goal of getting out in nature for an overnight backpacking trip four times a year. This is how he reconnects with himself and God. Us discussing what our goals are and why, as a couple, has been really helpful to our understanding each other in our marriage and understanding how we can help one another make time to pursue our goals.

My sister has a spiritual goal of listening to a sermon every morning on her drive to work. She also wanted to make herself feel more comfortable praying out loud with people when they needed it. These are small spiritual goals, but they have produced really amazing changes in her life. Now when she has a patient (she is in her last stage of clinicals as a family nurse practitioner) that gets bad news, or is scared or in grief, she notices the opportunity to offer prayer when it is appropriate. She would have never noticed those opportunities had she not set them as goals. My sister has called me to tell me stories of how she saw God use her in unbelievable ways by placing her in the right room at the right time where she can offer a spiritual hand to someone who feels like they are drowning. One patient told her thirty times over, "I don't know how I would have gotten through today if it wasn't for you."

My sister will tell you she has never been the extroverted type. Neither has she been the type of person to lead a Bible study or even pray out loud at a family dinner. Her faith has been very complex, personal, and quiet. Yet, here she is now feeling as though God has never been more real in her life, all because she set this goal to start offering to pray out loud when she felt the situation may be appropriate. What a small goal that has produced truly incredible moments in both her life and other people's lives.

My friend Scarlett has the goal of finding recipes off Pinterest once a month so she can make a treat for her three daughters that looks like it belongs in a magazine. She made cake pops recently and sent them to school and they felt so special. Maybe she would have made a treat once a month anyway? But now she has the goal to, and it helps her notice when she doesn't. It helps her to be more intentional in her parenting. Remember, once you write down your goal, you become 33 percent more likely to achieve it.

Habits of Successful People

In the book *Change Your Habits, Change Your Life,* Tom Corley writes about the habits that made average people self-made millionaires.[7] Some of his more interesting findings may shock you. For starters, *when* you start your day has a pretty high correlation to how successful you are. Nearly 50 percent of the people in Corley's research reported getting out of bed three hours before they had to be to work. I read *The Circle Maker* by Mark Batterson several years ago, and something he said changed my life. "The biggest difference between success and failure, both spiritually and occupationally, is your waking-up time on your alarm clock."[8]

This was when I started setting my alarm clock for 5 a.m. I realized I needed to intentionally pursue God. I wanted to make myself uncomfortable in that pursuit. The day was not going to give me time to suddenly have the deep fulfilling relationship with God I wanted. I needed to create it. This habit I started fundamentally changed my worship life almost immediately. It was a perpetual reminder I was in pursuit of Christ, and I wasn't asking him to meet me on my terms. I was willing to make myself uncomfortable. I have now moved my worship time to 4:30 a.m., because I like to sit and just chat with my husband uninterrupted by kids before I go to work for the day. We have coffee and we whisper.

I am trying to be intentional in my marriage and my spiritual life about prioritizing time together. I could sleep thirty more minutes, or I could connect with my husband and then have worship? For me, it is a no brainer.

Another habit of successful people is 88 percent of them said they spent thirty minutes a day reading for educational improvement. Reading is one of the best things you can do for your brain. (Look at you right now, setting yourself up to become a millionaire!) In an article by the founder of AllTop-Startups, Thomas Oppong, called "The Reading Brain (Why Your Brain Needs You to Read Every Day)" he writes, "As you read these words, your brain is decoding a series of abstract symbols and synthesizing the results into complex ideas."[9]

Reading allows your brain to thrive. It activates your thinking and coordinates deep processing of tons of stimuli. It connects so many different areas of your brain at once that it is really no wonder this one activity is so deeply ritualized by successful people. And at a time when we struggle to avoid mindlessly scrolling social media for longer than thirty minutes at a time, Oppong says reading even helps your brain improve the ability to be patient. It retrains your brain to focus on a single task at hand, during a time when we have probably never multitasked more.

A couple years ago, my friend Vimbo and I realized if we wanted to be experts in our fields, we had to read. We decided to try to read twelve books a year, one book a month, on the exact same topic. The idea being that within a year, we would know so much more about that topic and the different conversation pieces and leading speakers and writings on it than close to anyone else. I didn't finish twelve books, but I probably got through six books that first year all on the topic of communication. I never struggle to do a podcast interview or give a lecture now. Within that first year, I had doubled my content knowledge, and now every book I add to it just develops my expertise even more. Tomorrow, if you wanted to become a speaker or social influencer on a topic, all you have to do is start reading what people who already are impacting that space are saying. And then read it again and again. Twelve books later, in one year, where do you think you will be?

Corley says highly successful people have also incorporated a habit of spending fifteen to thirty minutes a day on focused thinking. This is the same research we find in Newberg and Waldman's *How God Changes the Brain* which discusses the benefits of meditation. Spending time just on thinking is probably one of the hardest things for me to incorporate. I love reading and I have learned to love to get up early because in both scenarios I am able to do something that makes me feel productive. As an Enneagram 3, producing is all I know.

Taking the time to allow my brain to just stop and be? That is harder for me. But the more research I read on meditation, the more I knew I had to incorporate this practice into my life. They have studied this with children, with medical professionals, and those with high-stress jobs: the benefits of quiet time to catch your breath and collect your thoughts is a really great tool for your brain. Take fifteen minutes a day in a quiet room and get to know yourself. Ask the questions you may not consciously be acknowledging. *Am I happy? Am I where I want to be? How are my relationships?*

Just like you would need to touch base with your romantic partner each day in order to keep your relational intimacy strong, checking in with yourself is extremely important, and most of us never do. I had a therapist once who told me to start putting my hand over my heart before bed and saying my name out loud with an affirmation. It sounds silly, but the first time I did it and said *Heather, you are worthy*, I cried. I had never acknowledged myself as a human being before. I had never tried to really speak intentionally to my brain. I do it all the time now, and it helps me stay in touch with my emotions and not let small things suppress until they become big things. Try it. Put your hand over your heart, say your name, and then ask yourself a question and speak an affirmation. Studies have shown hearing your own name is one of the strongest ways you can activate your brain.[10]

Strong communicators know this instinctively. They have the gift of walking into a room and making everyone feel like they know them. One small trick to doing this is simply using people's names when you are in conversation. "Wow, Gabby, tell me more about your family." Or "Interesting, Josh, what made you want to move here?"

Using people's names is a quick way to activate their brains. Every teacher knows one of the quickest ways to build rapport with your classroom is to memorize the names of your students. The research is so impactful, even people in vegetative states showed brain activation, even if only for a moment, when hearing their own names.[11]

The last habit in Corley's research I want to touch on is that successful people spend time with people who inspire them.

It will be exceedingly difficult to get where you want to go if you are the only person you know who wants to go anywhere. I have seen some of my brightest students end up in jail because they surrounded themselves with friends who weren't good for them. I have watched incredibly talented, sharp women end up with the wrong guy and get so stuck in the relationship that they lost sight of themselves. Your relationships are incredibly important to your success. Just like in a race, a runner is often only as fast as the competitors they have pushing them. You will run as hard and fast as whoever is running next to you. If you feel you are lacking inspiration, look for a friend who inspires you. In order to find happiness, relationships are key. We are intentional with our jobs, intentional about car purchases; we may even research the dog breed we'd like. And for whatever reason, all of us suddenly stop being picky when it comes to who we let into our lives? The best thing that may ever happen to our motivation is simply making good choices with our friends. Choose wisely.

Write out your own SMART goals here (specific, measurable, achievable, relevant, time-bound):

S-

M-

A-

R-

T-

Promise six to memorize:

"Little by little I will drive them out ahead of you, until you become fruitful and possess the land." (Exodus 23:30 Berean Study Bible)

DISCUSSION QUESTIONS

► What do you see that God could be doing right now little by little? (Remember I asked what do you *see*, not what do you *feel*.)

► What time do you set your alarm? Do you think it should be earlier? Why or why not?

► Name one person who inspires you that is a close friend. If you don't have one, what is one thing you can do to intentionally meet more inspirational people?

7

It's Your Turn
to Network

*The single greatest "people skill" is a highly developed
and authentic interest in the other person.*

BOB BURG

M Y DAD'S HEALTH WASN'T GREAT, and I felt like it
wasn't the time to move for a job. A month before I
was supposed to relocate, I went upstairs in my room and
yelled at God.

"If this is what you want from me, you have to make it clear,
or else I'm not going!"

I meant it. I didn't have peace anymore. I cried into my
pillow. A couple hours later, I walked downstairs and remembered to send a Facebook message to an old friend who I heard
was pastoring where I was supposed to be moving.

"I heard you are in Denver," I wrote. "I am supposed to move out there in a month. Would love to visit your church." I hadn't spoken to Pastor Andy Nash in about four years, but if I did move, I would need a church. May as well visit his.

Immediately, my phone vibrated.

"Can you call me?" he texted.

My initial thought was *Bro, I said I want to visit your church, not talk on the phone!* But I thought that was rude of me, so I dialed the number.

"Heather, this is really crazy, but did Seth ever finish school?" The last time I talked to Andy, about four years prior, my husband was in and out of school trying to finish his religion degree.

"Yes," I said.

"I am about to walk into a board meeting. I need an associate pastor and you messaged literally as I was about to walk inside. I felt like maybe this was a God thing? Can I call Seth tonight? Think he would be interested?"

I almost dropped the phone. As long as I have known my husband, he has felt like he was supposed to go into ministry. His brother Tyler died his senior year of high school, and though he went away to college, I don't know if he ever left his dorm room. He failed out, developed severe depression, and ended up back in and out of school for the next several years. As his mental health improved, he did finish his program, and then, the kid who had straight Fs his freshman year finished an MA program with all As. He wasn't dumb, he was depressed. But for years, he didn't know the difference.

He had applied for some pastoral jobs and nothing seemed to work out. It just wasn't his turn, I guess. He was pretty bitter with God about it. His whole life, even in his deepest depression, he felt he was supposed to go into ministry, yet when he finally pushed past all his fears and went back to school and graduated, he couldn't get a job. He ended up becoming an assistant principal and working with at-risk youth. He loved it. He made it his ministry. But he struggled with God. He felt very strongly while Tyler was sick that the Holy Spirit was calling him to build a church. But no church ever called.

Now here we were, years later, right after I had prayed for God to give me peace if he wanted me to make this move, and Andy Nash, someone I hadn't talked to in years, was asking Seth to come pastor with him in Denver. Oh, by the way, the church was located within five miles of my job and where we would be living. Within three weeks of that phone call, Seth was hired. What if the key to what comes next rests in the hands of someone you already know?

What Are the Odds?

There are 7.8 billion people in the world.[1] Finding the person who could change everything for you doesn't seem easy. Ask anyone who is single, over thirty, and looking for love. According to a 2017 *Daily Mail* (UK) article, if you are looking for love tomorrow, you have a dismal 1 in 562 chance of finding it if you just leave it to destiny. The article reported on research from the University of Bath, where scientists looked at the mathematical odds of finding your person. While leaving it to fate may only give you a 1 in 562 chance, the researchers did list things you could do to improve your odds. Simply talking to people while at the gym boosts your odds of finding love by 15 percent.

Researchers based their study on eighteen factors they believed were key to falling in love: desired age, location, and physical attractiveness all played into who would be a potential match. Researchers found that in the UK, out of a population of forty-seven million people, only 84,440 people are potential matches for any given person. This equals a 1 in 562 chance of finding love. Quoting mathematician Rachel Riley, the *Daily Mail* writes, "When it comes down to it, love really is a numbers game. Obviously, the more people you make the effort to meet, the higher your chances of romantic success. . . ." In a world of over 7.8 billion people, it is a wonder any of us ever meet the right person at the right time, for anything.[2]

It's a Small World After All

Stanley Milgram, an American social psychologist, conducted a somewhat revolutionary "small-world experiment." His research indicated beneath this great big world, there is actually a hidden smaller one characterized by short networks. It is from Milgram's work we get the term "six degrees of separation." Researchers were extremely interested in the concept of the growing interconnectedness in America. Milgram's study found people tended to be connected by three friendship links.

Malcolm Gladwell talks about Milgram in his book *The Tipping Point*.[3] Milgram mailed packets to 160 people living in Omaha, Nebraska. Each packet had the name and address of a stockbroker who lived in Massachusetts. The recipients were instructed to write their own name on the packet and then, trying to get it as close to the stockbroker as possible, send it to someone near Sharon, Massachusetts who may know the person. The goal was to get the packet to the stockbroker in as few steps as possible. While some may have had zero connection to Sharon, Massachusetts, they may have had a relative who lived closer than they did, and may know a guy who knows a guy.

Once the packet finally arrived in the mailbox of the stockbroker in Sharon, Massachusetts, Milgram then read the names of everyone who had handled the packet. Some of the packets reached the target destination in as little as two links, while others took as many as nine or ten. The average, however, of these chain packets took about five or six different exchanges of hands before the packet reached the stockbroker in Sharon, Massachusetts.[4] Keep in mind, this experiment was before social media and the internet, and the average American was still linked to a total stranger in about five or six steps. The answer to what you are looking for probably lies in the hands of someone you already know.

The study was shocking. Here were people from a smaller Midwestern area knowing people halfway across the country. Gladwell writes that when Milgram asked a friend how many tries he thought it would take to reach Sharon, Massachusetts, he estimated one hundred. After all, in the UK, there are only eighty-four thousand possible matches for every romantic connection, not to mention expecting to get an envelope to the house of a complete stranger. How in the big world we all know is this possible? The answer may surprise you. Gladwell puts it this way: "In the six degrees of separation, not all degrees are equal."[5]

What Milgram found should shock you even more than the concept of six degrees of separation. "Twenty-four letters reached the stockbroker at his home in Sharon, and of those, sixteen were given to him by the same person, a clothing merchant Milgram calls Mr. Jacobs." Gladwell goes on to say that half of all the packets making it to the stockbroker came from the same three people. "Six degrees of separation doesn't mean everyone is linked to everyone else in just six steps. It means that a very small number of people are linked to everyone else in a few steps, and the rest of us are linked to the world through those special few."[6]

One way Gladwell suggests you can test this concept is by making a list of the top forty people you would place in a friendship circle or connection for your life. This circle should not include people you know from work or your family. Look at the names and think about who was responsible for bringing you into contact with this relationship. What you will probably discover is you don't have forty random friends, you have forty people who are similarly connected to the same handful of people who keep bringing you in. Remove one or two names from that circle, and the entire circle disappears. Most of our relationships are in fact not random at all. They are brought to us by people Gladwell calls "connectors." Connectors are the people who go through life bringing others together.

Becoming a Connector

What if you aren't naturally the connector type? And according to Gladwell, there *is* a type. Is it possible for someone to become a connector who maybe isn't naturally wired that way? In an article from *Vault Guide to Networking*, former Wall Street vice president Miriam Salpeter says yes. According to Salpeter, forcing yourself to perform gestures that would allow you to build a relationship—even if it's not for anything specific in return—is how you begin to think like a connector. Connectors are driven not by actual networking as much as relationship. Offering to do favors for people, making introductions for them, is how connectors behave automatically. Thinking about how you might help others can help you forge a stronger network. She mentions other basic skills, like joining organizations, sitting next to a stranger, and asking them questions without feeling the need to talk too much about yourself. Learning to listen and seizing opportunities to meet people, even when it's not convenient. Looking for ways to stay connected after meeting someone and following up with them are all ways you can become a connector.[7] Connectors are networkers, but they aren't focused on advancement as much as meeting and knowing other people.

Why Networking Matters

It is estimated that 85 percent of all jobs are filled via networking.[8] According to career social tool LinkedIn, networking works better than outright applying for a job by 3:1. In an article by *Forbes* contributor Falon Fatemi, successful networkers: (1) spend time focusing on networking, (2) keep in touch with their network when things are going well, and (3) start casual conversations.[9] Of course, certain personality types may lend themselves more to networking than others, but as a communication professor, I truly believe it is a basic skill that can be learned. Social media has made the ability to stay connected even easier. As someone who is probably a naturally wired connector, a few tips I can give is to check in on other people: a simple text checking in with an old boss to ask how he likes his new job or commenting on a college friend's new podcast post to encourage them along. Even asking for favors can make people feel more connected to you.

Author Barry Davret talks about one strategy to forming a connection with someone you could use as a potential networker. The name for the theory came from an idea Ben Franklin reportedly used in order to turn enemies into friends. He wrote about it in his autobiography, and in 1969 researchers agreed Franklin's maxim was actually confirmed through evidence.[10] In chapter nine of Franklin's autobiography, he tells the story of trying to make a supporter out of an adversary. Franklin, in hoping to make himself more likable, did something that on face value doesn't make much sense: he asked the adversary for a favor—if he could borrow a rare book. The person who hadn't liked Franklin, upon receiving the request to send the book, agreed to send it, and Franklin borrowed it for a week before sending it back. Franklin claimed the next time he spoke to the person who he borrowed a book from, the person was much more civil. His philosophy was that when people do something nice for you, they are more inclined to do something else nice.[11] A way to be a connector would be to ask your neighbor if you can borrow a shovel. When we first moved into our new house, my husband, who is more introverted than I am, went and asked if he could borrow a ladder. A couple days later, while grilling in the driveway, he asked our neighbor Blake if he wanted a plate. Next thing I know, my introvert husband is shooting hoops with our neighbors. And it all started by asking for help.

Asking people for a favor is asking them to do something nice for you. Once they do it, cognitive dissonance tells them they did the nice thing because they must actually like you or have a good relationship with you, even if that was not true in the first case. Cognitive dissonance is something we talk about a lot in communication. It encapsulates the idea that our brains like to be in congruence with our actions, and if they are not, we often don't change our action, we change our belief. So, if I am a smoker, and I believe smoking will kill me, my inconsistency between my behavior and my belief are very distressing to my brain. Interestingly, what most of us will do is not lose the behavior, but the belief. It is not that we will suddenly stop smoking, it is that we no longer believe smoking will kill us. It's a fascinating phenomenon and apparently it can work to make yourself more likable. If you ask someone for a favor and they perform it, their brains don't know what to do with the inconsistency. Rather than stop the behavior (being nice to you), their brains will tell them the reason they are nice to you is because they do actually like you. This idea of asking people for favors in order to become more likable became known as the "Franklin Effect."

In my own experience, one of the best tools for networking is simply asking people for help. What if you just asked someone you admired if you could pick their brain for fifteen minutes about their career or area of study? Remember for connectors, the goal isn't necessarily networking as much as it is relationship. We shouldn't reach out to people disingenuously, but to tell someone you admire that you value their expertise and work ethic and would love to just Zoom for twenty minutes to ask them three questions could be a great way to connect. One of the best things you can do to build relationships is to let people help you.

Some of my best networking relationships formed from me asking them to speak to my students. Five out of ten times, the people who I asked to speak or Zoom into my lectures end up being more friendly with me on social media, and eventually emailing or direct messaging me about opportunities they think I could help them with. I kind of just stumbled on this wonderful networking strategy a few years ago, and it honestly has changed my life. I have genuine relationships with people who are so much more successful than me, and it's not because I helped them, it's because they were willing to help me.

We have to build genuine time into our day to check on our relationships. I think this is the most important aspect of networking. Not asking people for anything but simply engaging with people as they are. If checking in doesn't come naturally for you, start scheduling it. Schedule in fifteen minutes every day where you will spend time on social media or in text messages checking in on relationships you value.

Send people articles you find interesting, ask them for their opinion on a piece of content you are creating, and of course, simply ask them how they are, and how you can pray. I try to make a habit of doing this with almost everyone in my legitimate friendship circle at least every other month. Even if my short "I'm praying for you this morning," text doesn't strike a conversation, it reminds people that I care about them, and it reminds me to care about other people.

I read a *Huffington Post* article called "11 Ways Introverts Would Prefer to Start a Conversation" that gave some really great tips for how introverts may go about trying to be a connector and networker. One person said asking about someone's pets felt very natural for them and less invasive than small talk. Another person said it felt more comfortable for them to ask people questions related to books when they were trying to form a relationship with someone they found interesting. Another interviewee said it was easiest for them to bring up travel plans in order to strike up a conversation.[12] All of us will probably have different ways of approaching connection, and there is no right or wrong way. There is simply your way, and you have to do what feels natural and comfortable for you.

Social Media and Social Networking

When it comes to social media, I notice most people tend to use it as a form of broadcast. I am not saying this is an inappropriate use of the technology, but I do teach courses on social media and am always surprised by how many people seem genuinely stunned when I remind them social media is an inherently social system. The point is not simply to get on your soapbox and dole out wisdom, the point has to be to make social connections. If you want to improve your online networking ability, the first thing to do is to simply use it socially. Comment on other people's posts. Retweet and share their thoughts. The goal on social media isn't to create a slew of followers, it is to create a faithful community. Stop thinking you need ten thousand people to have a platform. What you need is even just one hundred people who have fully bought in to what you do.

I know social media strategists who make their entire living off five thousand followers, and I know social media influencers with one hundred thousand followers who aren't making a dime. The temptation of social media is to see numbers and think in terms of mass penetration. I always tell my students to stop looking at follower counts and start looking at engagement ratios. If you have one hundred thousand followers, but only one hundred of them are liking your content, it doesn't really matter that you have one hundred thousand followers. And if you ever need to sell them something, none of those people are buying it. But if you have one thousand followers, and every time you post you are getting one hundred likes, you are actually doing amazingly well at creating a community. Ten percent of your community is repeatedly showing up for your content. Typically, only 10 percent of your followers are even seeing your posts![13] Imagine if the person with one hundred thousand people could have 10 percent of their market constantly engaging? They would be going viral at every post. This is why slow growth can be better online than fast followers. The people who are coming to your site feel connected to your content, so when you have something to sell them, engaged people are going to be far more likely to want to buy it because they want to support you.

We have to stop thinking about social media as building followers and start thinking about how we can build community. Connectors don't need followers, they need relationships. Before you log online think, *people, people, people*. This is your networking chance to connect with other human beings. The best way to connect with people is to care about them, rather than try to convince them to care about you.

Doug

Once while flying from Denver to Orlando for a speaking engagement, I found my seatmate was a man in his midseventies named Doug. Before the plane even took off, Doug started chatting with me. He asked the usual questions, like what I did for a living, where I was headed, and where I was from, but somehow those answers led us into deeper conversation. I chatted with Doug the whole three-hour flight. He was also a public speaker, so we bonded quickly. He did corporate events training business leaders on cognitive performance coaching. At one point, Doug handed me two of his business cards.

"I always give two cards," he said as I tucked them both into my purse. "One for you, and one for you to give should you ever meet someone you think would benefit from having my card as well."

Doug was a connector. He was absolutely a networker, but he was also interested in me. He complimented my storytelling, asked me questions about my parents, and even reflected on my faith. Doug mentioned early on in our conversation that he wasn't a believer and didn't want to talk about God. He only set these boundaries because when he asked where I was going, I told him I was on my way to speak at a women's faith conference in Orlando.

I didn't feel the need to evangelize Doug. I enjoyed our experiencing one another as people, not numbers. Before the plane landed, he said, "A couple times now you have mentioned that you pray. Heather, what are you hoping to get out of your prayers?" he asked sincerely.

"Well," I said, "If you want, I can tell you a story about prayer."

"Try me," he answered. So I told him the story about how I had called off my engagement two months before the wedding. How I had prayed to God that if I wasn't supposed to marry this person for God to please make it clear, and within minutes of me saying "amen" my phone rang, and it was my fiancé, and he had called to break up with me. On the same night, several hours later, Seth, my now husband, called. I told Doug my story, and I said, "Doug, that's what I hope to get out of prayer: God's involvement."

"I see," he said. About thirty minutes later, as the plane landed and we were about to start the mad dash to collect our things, Doug, who three hours before had told me he was a Jewish agnostic, grabbed my hand and said, "Heather, I am going to think about your God."

Doug and I have stayed in touch. One time, my husband found a selfie on my phone of an old man holding a book. "Who is this?" Seth asked, confused. The man was Doug. He had bought one of the books I recommended on the plane, and he wanted to tell me he had finished reading it. One time, he took a speaking engagement in Denver and called asking me to check my schedule to see if we could meet for lunch. As I was writing this, I texted Doug. It struck me how badly coronavirus had hit Florida. I worried about my plane-ride friend, so I sent a networking text that felt a lot less like networking and a lot more like friendship.

"Thinking of you," I wrote. "I hope you are well amidst coronavirus."

"I have the beginning symptoms," he said. "Extreme fatigue, cough, but I'm still busy working."

"It's okay to rest, you know," I texted back.

"Heather," the message came through, and I could feel its weight, "I could use some of those prayers."

I was stunned. He remembered our exchange. My testimony had stuck with him six months later. The man who once told me, "I don't want to talk about religion," now asked me to pray. And so, I did. I typed out a prayer, and I committed Doug into the hands of a God who knew him far more than I did.

Networking is important. But it is not just your career that will need your networking—your faith will too. Doug made it clear when we sat down that he was not interested in my religion, and yet, after three hours of genuine human to human connection, it was Doug who asked me why I prayed.

And the reality is that sometimes, the closest to God someone else will ever get could be you. I believe God is looking for connections. God needs the priesthood of *all* believers, not to share their faith in hopes of a number, but to talk to people about their stories. Something I tell my students all the time is that people may not want your religion. They may not be interested in your theology. But they will listen if you simply tell them your story.

Networking, connecting with other human beings, is one of the greatest gifts awarded to humanity. Being willing to have a conversation with a stranger may make all the difference in what happens next in your life. God needs social Christians. Connection is incredibly important, especially when *It's Not Your Turn*.

> ### *Promise seven to memorize:*
> "As iron sharpens iron, so one person sharpens another." (Proverb 27:17)

DISCUSSION QUESTIONS

- ► What is the best conversation you have ever had with a stranger?

- ► Who are three people you can text to check up on today?

- ► Is it hard for you to connect with people? Why or why not?

- ► What can you do to be more intentional about fostering relationships?

8

It's Your Turn to Take a Second Look at Power

The Pew Research Center study showed that millennials had far more negative views of their generation compared with Generation Xers, baby boomers or other age groups. More than half of millennials, 59%, described their generation as "self-absorbed" while 49% said they were "wasteful" and 43% said they were "greedy."

THE GUARDIAN, SEPTEMBER 2015

I WOULD BE WARY OF ANY PERSON who is the perpetual hero in every story. Sometimes, it is hard to admit we are just as susceptible to our dark desires as we are to want to push toward the light. I don't think I realized for years how dangerous my prayers were. *God, give me a ministry; God, give*

me a book deal; God, please help me to have a bestseller; God, please help me to make money." Me, me, me, me, me.

I saw a tweet by Pastor Ronnie Martin (@ronniejmartin) that said, "If we're being honest, the desire to do 'big things for God' is a convenient way to mask our desire to do big things, God or no God." Can I be honest with you? It has only been recently I even fully realized my selfishness. I have spent the last couple years really coming to terms with my own desire to do "big things" and whether they were really about big things for God or big things for me. I dug deeper and discovered so many of my prayers were not just about promoting God, they were about promoting myself. One day, God whispered, "Heather, you should never ask for me to exalt you. Just ask for me to be with you." It felt like a punch in the gut.

I started running through all of my prayers, and, of course, I had prayers for others: I prayed for my husband's church to thrive, I prayed for my best friend to find a husband, I prayed for my sister to crush grad school, I prayed for the health of people I knew who were sick, but nearly every prayer for me was about what I needed God to do for me. How I needed him to expand me, promote me, sell me, grow me. When I felt the Holy Spirit say, "Heather, you never need to ask me to exalt you," I sobbed for a long time in a dark room. I didn't see myself as having any resemblance to Lucifer. I was a good person, and my dreams have always been wrapped up in ministry. Suddenly, I saw very clearly the line between me and the devil wasn't that far at all. I was not the first person to want to be exalted.

The Lucifer Effect

Isaiah 14:13 speaks of the fall of Lucifer as being prompted by the devil's thirst for power: "You said in your heart, 'I will ascend to the heavens; I will raise my throne above the stars of God.'"

It is Lucifer's desire for power, for exaltation, for recognition, for magnification that leads him to becoming the adversary, or accuser, of God. What he didn't count on was Jesus literally becoming the second Adam and redeeming all of humanity back to himself. Do your prayers reveal the heart of Jesus? Or the exaltation of Lucifer? It's a hard thing to admit. The heart of God is always surrounded by humility. I wasn't seeing that heart in my prayers for myself.

What kind of a God would allow himself to be hung on a cross? One who knows there is something greater than fear, shame, rejection, pain, and even death could ever take. And one who has tasted that there is a sweeter victory than thrones or crowns could ever provide. My friend Shannon Dingle has written from an honest perspective about that kind of power. A power that won't come with book deals or accolades. It's a power only built in the quiet spaces of our deepest pains.

Shannon's husband, while playing with their children at the beach last summer, was killed when a giant wave knocked him to the ground, breaking his neck. She tweeted, "Nothing in my life has linked God to safety. I've rarely been safe. But love is a magic stronger than safety. I want both. I always will. But being loved will always be more beautiful (albeit more vulnerable too) than being safe could ever offer" (@ShannonDingle).

Shannon, so honestly and vulnerably, proclaims the same words of Job 13:15: "Though he slay me, yet will I hope in him." Shannon has been changed through a deeply intimate relationship with God. And this is what Satan can never combat: a saint encircled by suffering who still proclaims God is good and love is a more powerful force than safety. (Ask Jesus.) These are the saints who cause Satan to lose dominion over this earth. The Jobs and the Shannons and maybe even you. This is what it means to colabor with Christ. This is what it means to defeat strongholds. This is what it means to be okay when it is not your turn. *And they'll know we are Christians by our love, by our love. Yes, they'll know we are Christians by our love.*

Christ comes to live as a man and die as a man, to reclaim the territories of both life and death. Romans 6:9 (NKJV) says, "Death no longer has dominion over him" because Jesus quite literally dies to conquer the territory of death so that we may have eternal life.

> And do not present your members *as* instruments of unrighteousness to sin, but present yourselves to God as being alive from the dead, and your members *as* instruments of righteousness to God. For sin shall not have dominion over you, for you are not under law but under grace. (Romans 6:13-14 NKJV)

Grace is the game changer.

That said, we should always remember the "Lucifer effect": even angels can become demons. The devil was once an angel whose love of God, and good, and morality became eroded by his love for *self*. And we are all at risk of doing the same if we are not vigilant. Self is the tiny seed that, if watered, chokes everything, even its host. The ally of God became the adversary because he would rather rule the gates of hell than be a servant in the kingdom of heaven. And I am not sure if he is alone. If selfishness is the gateway to Satan, we are living in a generation being overrun with devils.

In his book *The Lucifer Effect: Understanding How Good People Turn Evil,* Philip Zimbardo, creator of the notorious Stanford Prison Experiment, asserts that under the right circumstances, most people will do evil things.[1] He writes, "Most of us perceive Evil as an entity, a quality that is inherent in some people and not in others. Bad seeds ultimately produce bad fruits as their destinies unfold."[2] But what if evil is more about opportunity than moral fortitude? What if all that separates you from the bad guy is the level of emphasis you have placed on self? What if people aren't worlds apart, but degrees?

How selfish are you? Whenever someone calls you selfish, what they are really saying is you are paying more attention to your own wants or needs than anyone else's. Today, 71 percent of American adults think millennials are selfish.[3] Here is the thing: selfishness is not just immoral, it is the complete antithesis of who God is and what the gospel is about. It erodes the government of God. Selfishness is the seed of the Lucifer effect. When you refuse to clap for others while *It's Not Your Turn*, you water that seed just a little bit more.

Evil cannot just be suicide bombers and drug dealers. Evil, according to Zimbardo, presents itself multiple times a day through tiny—perhaps even unconscious—choices. Just like bad people are capable of being good to some people (serial killers can still have families), good people are capable of being bad to some people. Understanding this puts you back on the hook. Just because you aren't a bully to everyone, doesn't mean you aren't a bully to someone. I think understanding human behavior as complex helps us to be honest with ourselves about who we really are. What if you are both the good and the bad guy? Acknowledging this is how we begin to take ownership of our own behavior. This is the Lucifer effect. The idea that a little selfish seed—unchecked—can grow into a vicious garden.

Rats in the Cellar

Zimbardo says, "Most of us have the tendency both to overestimate the importance of dispositional qualities and to underestimate the importance of situational qualities when trying to understand the causes of other people's behavior."[4] In communication, we call this the "fundamental attribution error." When a waitress is rude, we attribute her behavior to who she is as a person, but when we are rude, we attribute that behavior to our having a bad day. C. S. Lewis describes this in *Mere Christianity* as having "rats in the cellar." In chapter twenty-nine, he writes what we often do as humans is feel badly about how we snap under stress or react when we are caught off guard. Then we say, "Well, that's not who I am. I just behaved that way because it was the wrong person, at the wrong time, on the wrong day." Lewis believes who we are in our most unguarded moments may be more of who we actually are than we'd like to admit. He calls this "rats in the cellar."

> If there are rats in a cellar you are most likely to see them if you go in very suddenly. But the suddenness does not create the rats: it only prevents them from hiding. In the same way the suddenness of the provocation does not make me an ill-tempered man: it only shows me what an ill-tempered man I am. The rats are always there in the cellar, but if you go in shouting and noisily, they will have taken cover before you switch on the light.[5]

Lewis goes on to say, what we are matters more than what we do. But who we are is often evidenced by what we do. What seeds of jealousy, seeds of vindictiveness, seeds of greed, of ego, of insensitivity, of racial bias are the rats in my cellar? And it doesn't matter if I only have to acknowledge they are there a few times a year. What matters is there are rats in my cellar, allowing me to pretend I am someone I am not. How do I go about confronting them, being honest about them, and removing them? I think, first, I have to turn a light on in the places where they like to hide.

Lewis continues:

> And this applies to my good actions too. How many of them were done for the right motive? How many for fear of public opinion, or a desire to show off? How many from a sort of obstinacy or sense of superiority which, in different circumstances, might equally have led to some unbelievably bad act? But I cannot, by direct

moral effort, give myself new motives. After the first few steps in the Christian life we realize that everything which really needs to be done in our souls can be done only by God.[6]

The truth is, we all have rats in our cellars. And who we are while it is not our turn determines who we will be when it is. Who we are when no one is watching may *be* who we actually are.

Power Corrupts the Brain

Power should come with a warning label. Every hit you try to get of it is like a temporary high. If it were a drug, you would need a prescription, and there would be a long list of side effects. You need to know this because when it is "your turn," that space of your life may come with great responsibility and an elevation of your power. This is all you have wanted for as long as you can remember, which is why the best thing God can do for you is let you sit and clap for other people. God knows what power is capable of. God knows angels can become demons.

While your powerless brain right now is confident you will use your space of privilege humbly, your powerful brain may corrupt your current innocence. Look at some of the horrors we have witnessed on Wall Street, some of the disappointments we have faced in politics, and the despair we have watched as over and over abuse comes to church. An article from the *Atlantic* exploring the work of Dacher Keltner, a psychology professor at UC Berkeley, examines decades' worth of research on the side effects of power.[7] Keltner found that under the influence of power, subjects acted as if they had undergone traumatic brain injuries.

Author Jerry Useem of the *Atlantic* writes that people who experienced power shifts became "more impulsive, less risk-aware, and, crucially, less adept at seeing things from other people's point of view."[8] Useem also references the work of neuroscientist Sukhvinder Obhi. As Obhi looked at the brains of both the powerful and the powerless, he found the brains of powerful people showed impairment on a specific form of neural processing that blocked one's ability to produce empathy. Once powerless people acquired power, their brains underwent changes, and they began to lose key functions they once possessed.

A 2006 study asked participants to draw the letter "E" on their foreheads where others could see it.[9] Those with powerful brains were far more likely to draw the "E" in a way that would look right to themselves, but backward to everyone who would read it. This is because, as Keltner explains, powerful people stop mimicking how other people experience things. Their brains stop doing an exercise quite common to the rest of us, which is putting yourself in the other person's shoes. Essentially, their brains tell them there is no time to care about anyone's shoes but their own. What is interesting about all this biologically is that the powerful brains weren't broken, but the scans revealed they had been anesthetized. As researchers continued various studies showing differences between the powerful and powerless brains, they started priming powerful participants with information to help them make conscious efforts to react to various tasks more empathetically. They still didn't. Even when they were told what mirroring other's emotions looked like, they were unable to actually do it themselves. It is no wonder Jesus, the only person to inherently deserve and have power, continually laid it down. Perhaps he wasn't just being humble, but wise. He was fully God and fully human, and he knew power was dangerous to his humanity, so he navigated it with great caution. We would be wise to do the same.

The Corruption of Power Systems

Communication professor Sonja Foss describes ideology as thought systems of representation employed by group members to make sense of the world around them.[10] A set of beliefs creates an ideology, and when that ideology is espoused among multiple members of a group, it becomes a system. It is our ideologies that allow a group to behave like a group. It is our espoused ideologies that align us with various members who think like us, and therefore behave like us with a goal of reproducing similar actions and responses. Some ideologies become privileged over others depending on the status of which the members of the ideology belong. When ideas are presented that are not part of the privileged system, those ideas tend to be repressed. "The result," Foss says, "is a dominant way of seeing the world. . . ." Hegemony occurs when there is a form of social control over ideologies where powerful groups of ideologies suppress the ideas of those with less power. When this happens, a group has the power to create "the enemy."[11] This is how Hitler was able to murder over six million Jews. It is how the Salem witch trials are even a part of our history. It's the Spanish Inquisition, the burning of Protestants at the stake during the Reformation, and the demonization of immigrants happening all across our borders today.

In a study assessing how religious and nonreligious students would view the others' arguments, it is no surprise both sides were primed to see the other with skepticism but missed the flaws in their own perspectives.[12] Our brain automatically roots for groups we have preference toward. This is why we trust one political party and oppose another. Our brain's desire to do this is so strong that even when we are placed in groups at total random, we begin to show preference for our own group members.

Newberg and Waldman detail the research of Princeton professor Susan Fiske, who says that if you want to operate with less bias toward others, you do that by not categorizing yourself.[13] Even our sincere desire to label ourselves as "Christian" can cause us to subconsciously place everyone who doesn't wear this label as an outsider. One study showed that within seconds of seeing someone from a different racial group, our brains illicit fear. Our brains do this naturally, so we have to prime them to be conscious of it.[14] We have to be careful as Christians to not create a hegemony that devalues outside perspectives. Religious ideology in the wrong selfish hands can wield "weapon Jesus" in dangerous ways. It is only when we as a church reject the need for power and establish an ideological commitment to service that we are truly able to restore the kingdom of God, on earth as it is in heaven.

When It's Not the Turn of an Entire Group of People

On May 25, 2020, police in Minneapolis arrested George Floyd, a Black forty-six-year-old man, after a convenience store called saying Floyd had just bought cigarettes with a counterfeit twenty-dollar bill. Video emerged of Floyd being pinned beneath the knee of Derek Chauvin for over eight minutes. For over eight minutes, a man suffocated, cried for his mother, and eventually died over twenty dollars. The world was outraged. Protests broke out across every city in America and abroad. The reality is this: no one would have watched a neighbor place a knee on the neck of a dog whimpering for eight minutes without intervening. Don't Black Lives Matter?

Floyd's death came on the heels of a video that surfaced of twenty-five-year-old Ahmaud Arbery, who was shot by a father and son pair while simply going for a jog. Gregory Mc-Michael and his son Travis, armed with a pistol and a shotgun, pursued Arbery with their pickup truck. They said he looked like someone who had been suspected of a series of break-ins in the neighborhood. Police have since determined there were never any reports filed regarding these alleged break-ins.[15] These men murdered Arbery, in broad daylight, in cold blood, and yet roamed free for over two months until video leaked of the shooting.

Breonna Taylor, a twenty-six-year-old nurse in Kentucky, was shot in March after police entered the wrong home on a drug investigation. By August, *O Magazine* started erecting billboards in Louisville asking for prosecutors to finally make an arrest after police knocked down Breonna's door in the middle of the night and killed her while she lay in her own bed. Breonna's family received a settlement of twelve million dollars, but still no arrests were made.[16]

While these murders and acts of police brutality and violence may be getting national coverage recently, Black people in this country have long been trees cut down in woods, with shouts for justice no one heard. And if a tree falls down, and no one chooses to see it, does it make a sound? Our country, which was built for freedom, was also built on the backs of people who were enslaved. For hundreds of years, Black people have had turn after turn stripped away from them based on corrupt systems of power. I am grateful many in America are finally listening to what Black people have been saying for decades.

During my PhD program, I had to create a portfolio of my work and defend it before a committee. At my defense, a woman on the committee asked me, "Are you saying racism is still so bad it impacts learning environments?" My dissertation explained how storytelling can be used to create connection between faculty and students, and at some point, I must have mentioned how important it is for students of color to see representation in their faculty. I said, "Well, I, a Black student, straightened my hair before coming today. I knew that even my natural hair would change how you, my educator, perceived me." I was met by silence.

One of the Black committee members spoke up: "Yes, we know racism can impact achievement. Let's move to the next question." He tried to change the direction. The woman interrupted him: "I don't think that's true. I think it's just related to poverty, but Black people are so loud that we only focus on their stories." Her follow up statement revealed her own bias, and she made my point for me. She dismissed the cries of racial injustice by Black people and felt comfortable doing so, at my own portfolio defense, in front of her Black colleague. None of what I said, or the research I showed, mattered. I was just being loud.

The truth is, I know as a woman of color that our country is so riddled with bias that just leaving my hair how it naturally grows out of my head may be the difference between whether or not I get a job. It's one thing if it is not my turn because sometimes doors don't open and life isn't fair, but it's a whole other conversation when it's not been your turn for generations and generations of people. I think this is why the gospel is so precious to so many people of color.

Job 35:9 reads, "People cry out under a load of oppression; they plead for relief from the arm of the powerful." The stories of liberation, the unjust persecution, the Israelites crying out to a God they sometimes worry isn't listening anymore: it's not abstract words in an ancient text for us. It's being seen, being known, and being heard by generations who walked before, not just spiritually, but physically carrying some of the same burdens. Exodus 3:9 says, "And now the cry of the Israelites has reached me, and I have seen the way the Egyptians are oppressing them." The God who delivered Israel from the arm of Pharaoh is the same God we will stand before at the end of time for judgment. The same God who said to Cain in Genesis 4:10, "Your brother's blood cries out to me from the ground" will call us to account for the blood we were too uncomfortable to confront on our own concrete. A powerful verse that always makes me stop in my tracks is Ezekiel 35:6 which says, "Since you did not hate bloodshed, bloodshed will pursue you." Oh, church, have we hated bloodshed?

The Responsibility of the Powerful and Privileged

I used to teach at a community college in Southwest Michigan. One of my students who never really laughed at my jokes (and honestly, I am not sure liked me that much), often sat in the back of the class. Suddenly, I had a strong feeling I needed to give her the money I had in my wallet. I had twenty dollars. I felt this was the Holy Spirit, but this was also a secular school, and it didn't seem appropriate for me as a professor to shell out cash to unsuspecting students.

Sorry, God, wrong number, I thought to myself.

Class ended, and as the students exited, I heard the Spirit whisper, "Heather, if you can't be faithful with small turns, how can I give you big ones?"

Now, I was guilt ridden and started running around the parking lot looking for this student. I couldn't find her. That was the Friday before spring break, and I had a trip to Mexico planned with my husband. We had a great time, and I didn't think of my student or the twenty dollars once. I got back to school a week later, and what would you know, the second she walks into the classroom I hear the voice again (by the way, it wasn't like this deep Morgan Freeman-God voice. The voice just sounded like my own thoughts, which is why I often ignored it) say, "Heather, give her the money in your wallet." This time it's forty dollars—so God doubled it.

I called her to me after class, and I totally fumbled and mumbled something like, "This is super weird. I am a Christian, and I feel like God is telling me to give you this forty dollars. I have no idea why, this money isn't from me, it is between you and God, and I just want to be faithful." I handed her the cash and was ready to book it out of there when she started sobbing. I did not know she lived in a halfway house through a church lady. I did not know she had a six-month-old baby. I did not know she had been a victim of sexual abuse and spent years in foster care. And I did not know she was only nineteen years old.

"Before I came into this class," she said, "I did something I haven't done in a long time. I prayed. I prayed that somehow, if God were listening, he would give me enough money to buy a box of diapers." And here I was, an hour later, handing her forty dollars. She went on to tell me she hadn't prayed in years. God never seemed to listen anyway.

"I cannot believe this is happening right now," she kept repeating. We became like family after that day. Her daughter, who was six months old then, is a hilarious, active, precious little girl now. My student has her own place, a car, and a job. I even introduced her to another student, and they have been dating for over three years. All of this happened because I stepped into the uncomfortable space of doing what I believed God told me to do, of listening to that still small voice and acting on it.

There are people right now who feel abandoned by God, and it's not because God is not answering their prayers, it's because he is prompting us to colabor with him toward justice, and turn making, and door opening, and we don't. I no longer see God as distant, and I no longer see my choices as inconsequential. God stopped my life all because a teenage girl in a halfway house needed money for diapers. God heard her, and I heard God, and it changed both of us forever. It was my responsibility to create a turn for my student. I was the hands closest to the door God could knock on. I do not know what would have happened had I ignored that responsibility. I don't even want to think about it.

We all have different levels of power in our various spheres. We all have varying degrees of privilege. How we use the turns God gives us to create spaces and turns for others is important work. Hebrews 13:16 (ESV) says, "Do not neglect to do good and to share what you have, for such sacrifices are pleasing to God." Philippians 2:4 (ESV) says, "Let each of you look not only to his own interests, but also to the interests of others." 1 John 3:17 (ESV) says, "But if anyone has the world's goods and sees his brother in need, yet closes his heart against him, how does God's love abide in him?"

Maybe it isn't your turn today, but here is what I can tell you, and here is why who you are when it is not your turn is more important than who you will be when it is: should seasons change and doors open, there will be a responsibility that comes with your new position. Whenever you get to where you are going, you must go back for those who are still where you came from. You have to create space for them.

Because sometimes someone else's turn depends on what you did with yours.

Promise eight to memorize:

"Blessed are the meek, for they will inherit the earth."
(Matthew 5:5)

DISCUSSION QUESTIONS

► Is there an area of life you could be less selfish in?

► How has wearing the Christian label created an "us"
  vs. "them" mentality for you?

► Are you concerned at all about who you could be should
  you get power? What could you do now to help your
  brain stay humble later?

9

It's Your Turn to Find Community

There is a peculiar gratification in receiving congratulations from one's squadron for a victory in the air. It is worth more to a pilot than the applause of the whole outside world.

EDDIE RICKENBACKER

I READ A TWEET BY EVANGELIST Ty Gibson (@tyfgibson) that said, "We live in a post-Christian culture. The faster we grasp that, the better. Even the Christianity that occupies the Christian space is largely American nationalism posing as Christianity. And yet, this is a prime opportunity to reframe God for a new generation of spiritual exiles." As we talk about Christian community, I hope we can talk honestly about the good and the bad. The beautiful parts, and the broken pieces. Because the church contains both.

My husband and I have worked in ministry environments that were hardly healthy. Ministry can be this exceptional experience where you see God part waters. But it can also be a painful environment. Every pastor's family knows that. According to the Schaeffer Institute, 70 percent of pastors are fighting depression. Seventy-one percent are burned out. Eighty percent believe pastoral ministry has negatively affected their families. Seventy percent say they don't have a close friend.[1] How do we create community where congregants and pastors are seen and safe?

I desperately want to bring people into church community, but I also want the people I bring in to be safe when they come. I work with young adults, and I hear it over and over and over how they have horrible experiences in churches where they did not feel safe. And I know that I, as an adult woman, have had experiences where I felt unsafe at church. I have had experiences where I felt like people were out to get me or my husband, and it all feels especially twisted when it's done in the name of Jesus Christ and because "the Spirit has just told them they had to correct you." It is my deepest desire for Christians to be leaders in building community. So, I am not going anywhere. I will never leave the church because, while I know how painful the church can be, I also know how life-changing it can be. I have friends at church who have literally become like family. I think of my friend Tiffany who radiates love and peace everywhere she goes. She has challenged me, and loved me, and believed in me when I couldn't. I think of church friends when we moved to Denver, on our first anniversary without familial support, demanded they watch our kids while we hiked through Boulder.

There are so many amazing Christians who have allowed me to see glimpses of God himself. But there are also negative experiences that prevent people from seeing God clearly. The cross is to be carried, not weaponized. I am so sorry if you have met "weapon Jesus"—Christians who have used prayer as manipulation, saying things like, "I'll pray for you," because they don't agree with you politically. Some people will use even the sanctity of prayer to create a spiritual chasm where one appears superior and the other is inferior and in need of divine intervention. Prayer should never be used manipulatively. It's spiritual abuse. And yet students will tell me stories about how prayer now makes them feel uncomfortable because of how it was wielded in toxic hands. I once had a boss sit me down in his office and totally berate me. His words were not just unkind, they were inappropriate. And then you know what he did? He said, "Let's pray." He insulted me on many levels, called into question my faith commitment, and then set himself up as a spiritual authority over me through prayer. I have since had to learn to be able to say, "No, thank you," when someone is trying to weaponize prayer. As my husband says, "Every Christian knows what it feels like to be kissed by Judas. And we are lucky to have a God who knows what it feels like too."

I used to think Christianity and the faith community were going to be this ideal place where everyone supported each other, clapped for one another, and dreamed together. Some places are, but we are liars if we don't admit some places are not. I want to do everything in my own power to create communities where people feel loved, seen, and supported, and I mean *all* people. Not just the ones who look and talk like us. In Luke 6:32-36, we see Jesus drop his mic on probably 90 percent of us:

> If you love those who love you, what credit is that to you? Even sinners love those who love them. And if you do good to those who are good to you, what credit is that to you? Even sinners do that. And if you lend to those from whom you expect repayment, what credit is that to you? Even sinners lend to sinners, expecting to be repaid in full. But love your enemies, do good to them, and lend to them without expecting to get anything back. Then your reward will be great, and you will be children of the Most High, because he is kind to the ungrateful and wicked. Be merciful, just as your Father is merciful.

And yet, for a country founded on freedom *of* (not *from*) religion, Christians seem to really struggle with loving and clapping for people who are different. Lawyer and Georgian politician Stacey Abrams tweeted in October of 2018, "My faith should never be used as a sword to strike down another community. It should always be a shield to protect." May our faith be a shield.

Do Christians Love Their Enemies?

A Baylor study found more than half of Americans are intolerant of other faiths, and since 2001, religious hate crimes have only grown. In 2017, religious hate crimes were second only to race hate crimes, according to the FBI.[2] What did Luke tell us Jesus said again?

A 2017 article reported atheists were more generous toward Christians than Christians were toward them.[3] A Barna article on why millennials leave church says, "Millennials who are opting out of church cite three factors with equal weight in their decision: 35 percent cite the church's irrelevance, hypocrisy, and the moral failures of its leaders as reasons to check out of church altogether. In addition, two out of ten unchurched millennials say they feel God is missing in church, and one out of ten senses that legitimate doubt is prohibited, starting at the front door."[4]

One of the top three factors millennials cite for leaving the church is what they believe to be moral failures. And the truth is, many of those who have left Christianity haven't left God, they have left church. Some 82 percent of Americans believe in a higher power[5] (which I think should be talked about more than a war on religion), and according to Pew, 72 percent of Americans believe in heaven.[6] A common belief among Christians today is that we are under attack for our religious freedom, but in reality, after the 2016 election, anti-Muslim attacks rose by 19 percent and exceeded even anti-Muslim attacks in 2001 after 9/11.[7] Just during Donald Trump's presidency, White nationalism has risen by 55 percent according to the Southern Poverty Law Center.[8] My friend and Syrian activist Ala'a Basetneh did a guest lecture for my students, and I noticed she wasn't wearing her head scarf. "I can't wear it anymore," she said. "I am no longer free to express my religious beliefs in this country because I keep getting threatened when I do. Someone even pulled a knife on me once." She spoke to my classroom full of Christian students. You could have dropped a pin and heard it; the room fell so silent.

Is it others' freedom of religion we are concerned about, or just our personal freedom of religion? Are we ready to love our Muslim brothers and sisters and carry their burdens? Are we loving our Black brothers and sisters who are fighting for their lives? How about our queer neighbors? Or are we hashtagging religious persecution when it happens to the Christian church while ignoring the persecution within our own pews? Do Christians truly want freedom of religion for *all* people? In my opinion, I think we have to. I think freedom for my neighbor secures freedom for myself. And the people we think are different from us aren't that different at all.

Newberg and Waldman in *Words Can Change Your Brain* found that 90 percent of the time when we talk about God, we aren't even expressing the same concept. Abstract concepts can also be sources of miscommunication and conflict because we rarely explain to others what these complex terms mean to us. Instead, we make the mistake of assuming other people share the same meanings we have imposed on words.

> Let's take, for example, the word "God." In our research, we queried thousands of people ... and discovered 90 percent of the respondents had definitions differing significantly from everyone else. Even people who came from the same religious or spiritual background had

fundamentally unique perceptions of what the word means. And for the most part, they never realized the person they were talking to about God had something entirely different in mind.[9]

But what if the God we feel angry about others rejecting is a God we ourselves would reject? Something we do in rhetorical analysis is "cluster criticism." Essentially, you search a manuscript or speech and identify the key words, and then write down the words clustered around those key terms. The idea is that words surrounding key terms can tell you more about how the writer or speaker views a subject that may be subconscious to them.

For example, Sonja Foss analyzed a speech by Donald Trump and identified the terms of country, president, and politicians.[10] The terms that clustered around *country* were *wrong direction, problems, broken, serious trouble,* and *total disarray*. These cluster words help a reader explain what exactly is making them feel a certain way about the terms they hear or read.

Newberg and Waldman asked participants to describe what faith-based or spiritual experiences made a deep impact on their lives. While looking at the key words of people's spiritual experiences, Newberg and Waldman didn't find any significant discoveries. Of the nearly 5,500 words people used to talk about their spiritual experiences, no common terms were used. For example, "God" was only mentioned 18 percent of the time, and "Jesus" less than 4 percent. Their conclusion? True religious and spiritual experiences are unique, which means God must deal with all of us very differently. And yet, we try hard to code him using terms that may be too narrow for everyone to agree on.

What if God is bigger than the box most Christians want him to stay in? And this big God can't be fully explained or even fully identified. The only common denominator Newberg and Waldman found in their participants was the idea that these spiritual experiences were good for them and had positively impacted their lives. That goodness could not really be quantified though—which sounds a lot like God to me.

Are Christians Happy?

I wanted to do some research on whether Christians are actually happier than secular people—and would being happier make us more likely to clap for others? According to a Pew research article, actively religious people are more likely than those unaffiliated with religion to describe themselves as happy.[11] In the United States, 36 percent of actively religious people said they were very happy compared to the 25 percent of nonreligious ones. Newberg and Waldman say, "Regular church attendance will lower your blood pressure,"[12] and while it may not be in drastic amounts, it shouldn't be downplayed that there are benefits to a healthy religious life. Those who attended weekly religious services were significantly less likely to have a stroke, and this statistic didn't show a preference based on denomination. Faith communities are good for our health and do make us happier. In fact, a 2019 *National Geographic* article found the evangelical denomination of Seventh-day Adventists are some of the longest living people in the world.[13] On the downside, religious struggle where we worry that we are being punished by God can significantly shorten one's life.[14] Again, I think it is incredibly important that happy Christians who are willing to clap for others do their due diligence to preach a loving God around the world because belief in an angry God is literally shortening our lifespans.

Jenny Santi, author of *The Giving Way to Happiness: Stories and Science Behind the Life-Changing Power of Giving,*[15] says when we give to one another, it activates the same pleasure sensors in our brain as food, money, and sex.[16] Helping other people and living an "other-centered" life makes you feel better. As Christians, we are called to serve, and serving doesn't just help someone else, it's the gift that keeps giving because service helps us too. Altruism is hardwired in our brains, and it makes us feel good. Helping and serving others may be what makes Christians so happy. In short, being a Christian is good for you.

Our Relationships with Each Other Matter

How we treat each other matters. It matters to God, and it matters to our own physical health. Our close relationships to others are one of the biggest influences on our lives. I have a best friend at work, and it turns out that friendship is a predictor of whether or not I feel connected to my job. In a Gallup study, only 30 percent of employees reported having a best friend at work, but those who do are seven times more likely to be engaged with their job. They submit higher quality work, have a higher well-being, and are even less likely to get injured. In contrast, those without tight work-place relationships have a dismal one in twelve chance of being engaged.[17]

Basically, the best thing you can do as a boss (or a church leader) is foster a climate in which workers feel connected to one another. When I feel safe where I work, I am more likely to be productive. I am not worrying about who is trying to take credit for what I am producing, who is trying to steal my job, or who is out to get me. I feel a sense of community; therefore, all the energy I would have spent worrying about social threats are able to go to my being a productive employee or member. We have to consciously create families, workplaces, communities, churches, and a country where people feel safe and valued. It turns out, someone else's well-being indeed promotes your own self-interest. Human beings are wired to be socially reliant creatures. One member's downfall will erode the entire group dynamic.

Brené Brown in her famous TED Talk on vulnerability says it best: "Connection is why we are here."[18] Connection brings meaning and purpose, and it turns out that mental health is connected to physical health. Your relationships are not just good for swapping memes, they are actually making you live a happier life. This is why you should fight for those positive relationships within your church communities, and why I have too.

A team of researchers wanted to study the impact of relationships, and so they brought forty-two married couples to the hospital for an experiment where several small wounds were made on their arms. A device then measured the rate the wounds were healing. When couples had reported having some type of hostility in their relationship it took almost twice as long for the physical wounds to heal.[19] Toxic people are not just hurting you emotionally, they could also be hindering you physically. We need positive relationships, and you need positive spiritual friendships, especially while you wait for your turn.

The definition of friendship as crafted by Dr. Joseph DeVito says, "Friendship is a relationship between two interdependent people that is mutually productive and characterized by mutual positive regard."[20]

Press pause. Did you read that? In order for your relationships to be friendships, they have to be *both* mutually *productive* and experience *mutual* positive regard.

Our friends should make us better people. Our churches should make us better people. They should enhance our productivity as human beings, and they should make us see ourselves in a more positive light. Want to know traits of a productive person? It is someone who is inspirational, focused, and a problem solver. If your friends are keeping you unfocused, unmotivated, and uninspired, they are literally the opposite of the friendship definition.

Friendship is what helps you thrive. The saying goes, if ever you are the smartest person in your room, find a new room. If ever you are the most spiritual person in your group of friends, you may want to make more friends. If you are the only person clapping for others at your church, you need to bring in people who will join you, or find a church where you can truly thrive as an encourager to your peers. Our relationships are vitally important to how we live our lives, and rather than ask why young people keep leaving, maybe we should ask why they should stay? What are we actively doing to provide an environment where young adults feel clapped for?

One characteristic of millennials is we want change in our churches. We don't do things just because that's how things have always been done. We question tradition, we are skeptical of titles, and we believe each individual should be seen, known, and heard.

Research from the Barna Group says only 4 percent of Generation Z members have a biblical worldview.[21] In fact, the Barna Group says:

- ► Only two in ten Americans under the age of thirty believe attending a church is important or worthwhile (an all-time low).

- ► 59 percent of millennials raised in a church have dropped out.

- ► 35 percent of millennials have an antichurch stance, believing the church does more harm than good.

- ► Millennials are the least likely age group of anyone to attend church (by far).

Millennials are the most educated generation to date. Roughly 34 percent of us have a college degree.[22] This is a generation of people who were taught to ask questions. And so, we do. We ask questions of our government, we ask questions of our bosses, and we ask questions of our church. Millennials are the most impatient generation this world has ever seen and studies are now showing millennials will click off a webpage if it takes more than three seconds to load.[23] This could mean they aren't going to wait for you to decide you will make room for them. What are we doing to make young people in our churches feel mentored and seen? We have to clap for each other, young and old.

Who Are You Next To?

People tend to pick friends that are similar to them. The proximity rule of close relationships says like attracts like. This is key because you may hear people say, "Oh that's just my friend. They are a little racist, sexist, or homophobic, but I'm not." Or "That's just my friend. They are a little rude, don't mind them." Unfortunately, it probably *is* a lot like you because like tends to attract like. When Grandma told you "birds of a feather flock together," she wasn't wrong. Want to be a kinder, better person? Make a kind and loving friend. Want to be successful? Hang out with successful people. Think you are a Christian? Do you have any deeply spiritual Christian friends? You need them because like attracts like.

So yes, friendships are crucial, and relationships are vital to your health and well-being, but it's important to be cautious with who you let into your intimate circle. Gallup found in their fifty-year-long study of well-being, the single biggest predictor of what leads to higher well-being is not *what* you are doing but *who* you are with. Likewise, a twenty-year Harvard study found your physical health is determined more by relationships than the food you eat, the exercise program you are on, and the genes you inherited.[24] Your community is more detrimental to your health then your grandpa's mental illness, that Big Mac you just slammed, *and* the workout you skipped. Stop spending hours at the gym and then hanging out in a community that brings drama and problems. Your social ties don't just reveal a lot about who you are, they directly impact who you will become. If you don't have any non-Christian friends, you have totally missed the point of sharing good news and living out your faith. But if you don't have a community of people with whom you can experience God, there's a flag on the play.

I have intentionally lost many of my friends, some of them Christians and some of them not, because I understand it matters who I allow into my life. I understand the people who I call on bad days are the same people who create my good days. I like to think I am a kind, loving, service-oriented woman, and I can attribute that to the beautiful people I call friends. I hope you can say that about your friends as well.

I need Christian community. It's what convicts me of my selfish heart, reminds me of my human nature, and inspires me to dig deeper in my devotional time. You may not think you need the church to help you be more spiritual. Perhaps you are rock solid in your faith and a prudent study of Scripture. I am not asking you to go to church because the building makes you holy. I am asking you to find a church because the community will keep you healthy. I'm not telling young adults they need church. I am telling young adults why the church needs them.

> ### *Promise nine to memorize:*
>
> "Those who accept my commandments and obey them are the ones who love me. And because they love me, my Father will love them. And I will love them and reveal myself to each of them." (John 14:21 NLT)

DISCUSSION QUESTIONS

► What do your relationships say about who you are?

► Religious people are more likely to describe themselves as happy. Are you happy? Is that happiness attached to your career or your relationships?

► Ninety percent of the time when we talk about God, our brains are producing different understandings of who God is. Share a story of a personal experience where God revealed a piece of his character to you.

10

It's Your Turn to Re-envision God

The best fighter is never anger.

LAO TZU

W HEN I WAS FIFTEEN, I wrote my first novel. I never published it, but it told the story of a Black girl named Samantha Higgins. I think I was navigating my own Blackness at the time and was able to pour into those pages exaggerated versions of what I was experiencing in my real life. There is one scene from my book I'd like to share where Sam gets made fun of for being dark-skinned at school and she runs home to her mother:

It was in my first-grade year that I truly faced my own color. Macy at the supermarket was no longer just Macy; she was

"Macy the White lady." And George at the parking garage was no longer George; he was "George, dark like me."

My mother, however, always discouraged this kind of behavior. "People are people, Samantha," she'd always say when I referred to the "White lady" or "George like me," "If you have a problem with their skin, take it up with God. After all, he's the one who made them."

But my eyes had been torn. I didn't see people, I saw skin. I saw ebony and ivory. I saw curly hair and straight hair, big lips and no lips, pitch black and brown. I saw pink lace, and hand-me-down plaid. I saw everything I thought people were without ever seeing the person. It was after one of mother's racial equality speeches in third grade that I finally had the courage to ask her, "What color is God then?"

This was a question that always plagued me. I didn't know if God would pity me too if he were light brown like Rosy. And I didn't know if God would understand my tears if he were "God, the White man."

"God," my mother began, "is bright yellow when the sun rises," she knelt down and whispered in my ear. "And when we go to bed, he's gray like the moon. He's blue when the wind blows and one day when you're in love he'll be as red as roses and as bright as your smile." She kissed my cheek and squeezed my body so tight it became hard to breathe.

"And what color is he when I'm crying?" I asked. A few tears ran down my cheeks and onto my mother's flawless curls.

"Oh baby, when you're crying," she said, searching my eyes now, "He's clear, so that you can't see him holding you."

How Do You See God?

I think I used to see God working in my life very linearly. I kind of saw God as someone who coached me along my journey, and if I screwed up too badly, he would replace me with someone better. Then one day, while reading Lewis's *Mere Christianity*, I realized God is timeless. God was not just going through life linearly with me, God also saw my life from above the timeline. God had chosen to walk with me, even though he knew I would mess it all up at various points along the way. I think I saw God as someone I had to keep going up the mountain to find. Then I realized God, seeing all of my timeline, had already chosen to come down the mountain to me. I saw God as angry, or at least someone willing to get angry quickly if you didn't keep him happy. I didn't realize God is accessible. God seeks, God finds, and God knocks. You don't have to find God because the story is God finds you. You just have to open the door and let him in. It's your turn.

How Jealousy Blurs Our Vision

I have a blog that I run with my best friend called *I'm That Wife*. We blogged every week for three years and had nine thousand followers. I was tired. I was tired of watching girls posting sexy photos on Instagram blast the growth charts. I was tired of seeing posts like "napping hehe" get more hits than my actual think piece. I was tired of watching people on their public platform be nothing like the private people I knew in person. I started asking God why he wasn't policing them better? Some of us were out here praying and fasting and others were feasting.

I didn't want to judge, but things got pretty judge-y. My jealousy blurred my vision. I saw myself as someone who had gone up the mountain, and I became resentful when God didn't say thank you. All I saw was myself and how hard I tried to get God to notice me.

Ultimately, this belief that God was holding out on me hindered everything I thought about who God was and who I was in relation to him. It also destroyed how I was able to see other people. I didn't see people; I saw competition. And God was the guy at the top of the mountain watching to see which of us would play the part well enough to reach him first. What happens when we start to want something more than we want God is that God suddenly starts to feel distant. That thing we want gets closer and closer on our hearts, and God feels further and further away. But God didn't move—we moved God.

I learned to say *It's Not Your Turn*. I prayed I could genu-inely clap for someone who had what I didn't. I prayed I could trust God to judge his children so I wouldn't have to. I decided who I was while it wasn't my turn would determine who I could be when it was.

In a 2015 study seeking to find a connection between how jealousy impacts empathy, scholar Chunliang Feng at Shenzhen University in China had participants perform a perceptual challenge.[1] They looked at one hundred dots distributed across a computer and had to say whether there were more dots on the right or left sides. Researchers told the twenty-two participants this challenge had been completed by 650 people, and then informed them that based on their scores, they were only a two-star player. Next, researchers told them some of the one-star and three-star players had to undergo a sensory test requiring their faces to be injected with a needle. They watched fake videos of the needle injections, and then were asked how much pain they thought the players had been in who had to undergo this horrific process. The players said they felt equal amounts of empathy for both the one-star and three-star players. But the brain activity of the participants said they were lying.

When they saw the videos of one-star players they believed had performed worse than them, their brains appeared to genuinely reveal pain and distress for the player. But when they saw the video of the three-star players, their brains went quiet. They weren't able to produce as much empathy for people who they perceived to be doing better than them at a stupid sensory test.

This is why jealousy is a root that must be worked through immediately. No matter what you tell yourself, it dampens your brain's ability to feel for people who are doing better than you, and that happens when you can visibly see they are undergoing a painful process. Imagine when you can't see them struggling—how does your brain process their success? We aren't naturally wired to be supportive of the people who are cooler or richer or prettier than us. This is a conscious choice we have to train ourselves to make. When I first made this choice for myself, I didn't announce it to anyone. That may be the very reason why I could actually follow through.

Deciding to Do Things Differently

How many times have you told people you would start going to the gym because you wanted to be seen as the type of person who worked out? What about committing to a donation or mission trip because it's important to you that people see you as selfless? Or how about announcing a tithing decision because you know that's what good Christians will do? When I decided I wanted to stop hindering my own prayers and confront my jealousy, it wasn't because I was trying to be a good person. It was because I knew I wasn't, and that may be what helped me actually do it.

An article by social psychologist Heidi Grant Halvorson explains that promises, or announcing our intentions to others on some type of behavior modification, only work if we are trying to change our behavior because of the behavior's sake.[2] But when we make announcements or promises to others based on how we want them to see us, it might make us less likely to do it. When we struggle to feel good about ourselves based on images we would like to live up to, it can make us more likely to promote those intentions to others.

A 2009 study in the *Journal of Psychological Science* found when we announce our intentions to others, we feel rewarded just by announcing them. This very act of telling people how we plan to change makes us less likely to feel the need to change. Halvorson says, "Ironically, the more important the aspect of your identity is to you, the less likely you are to go through with it."[3]

To study this, undergraduate psychology majors were asked to write down their study intentions for the next week. They wrote statements like, "I intend to work harder at statistics" or "I will read the chapters more thoroughly." For half of the participants, an experimenter read their statements. The other half were told their intentions were not part of the experiment and would be thrown out. Seven days later, the participants were asked who had actually followed up on and went through with the intentions. The group having their intentions read aloud were 30 percent less likely to have gone through with them. The sheer rush of having someone else— even a stranger—validate their intentions by hearing them made them less likely to actually do them. This should make all of us pause for a second and think through the intentions we post online, especially with culturally relevant topics like the Black Lives Matter movement. Our mere posting a hashtag may make us feel like we are off the hook for advocacy. How much of our social justice is performative?

I read a quote by J. I. Packer that said, "Any Christian worth his salt will read his Bible cover to cover every year."[4] I never told anyone this quote had affected me, but I did start, quietly, trying to read through my Bible. I made a promise to myself and I did it, not because I wanted to be perceived a certain way but, because I believed Scripture was important and had never read it in its entirety. For the last ten years, I have read the Bible cover to cover every single year, and I will continue for the next ten. I didn't do it for J. I. Packer though. I did it for me. What quiet promise do you need to make to yourself right now? Write it down, right here below, and let it be a secret between you and these pages. How should we start to be better people? *Quietly.*

How Scripture Changes Us

About five years ago, I made a pact with myself that I wouldn't check the notifications on my phone unless I had checked for a notification from Scripture first. I won't get on social media unless I have had worship and alone time with just my Bible and God. For anyone trying to read through your Bible, I read five chapters a day and finish the entire Bible about every nine months. No special reading program—just Genesis 1-5, and then the next day Genesis 6-10, rinse, and repeat until you hit Revelation. This is great because I'm only stuck in a tough zone like Leviticus for a week (you can get through anything if you only have to stay there a week). Interestingly, after you read the Bible seven or eight times, even books like Leviticus start to show you beauty and symbolism. The more I read the Bible for myself, the more my idea of who God was started to change. I had always seen God as someone we said was loving but who would also zap me if I stepped out of line. Who prodded me up the mountain, but who would replace me with one of my competitors if I didn't get it all right. I told people God was forgiving, but I would go months struggling under the weight of what I was sure was his condemnation. Honestly, I think this is what a lot of Christians struggle with: worshiping an angry God.

The research agrees with me. Seventy-two percent of Christians see God as critical, authoritarian, and distant, and I was part of the majority.[5] If I am honest, most of the religious choices I made before reading the Bible for myself were out of fear rather than love. I lived with these inconsistent ideas of who God was for years, but the more I read Scripture for myself and stopped having other people control its narrative, the more I started to feel like God was someone I could trust. My friend evangelist David Asscherick (@dasscherick) says the entire Bible can be boiled down to the story of how God keeps his promise to one person, Abraham. I had never heard Scripture described that way, and I had never heard God described that way. God, not as the wrathful, vengeful, angry God I had heard about in churches and had conjured up for myself, but, as the promise keeper. And Scripture, not as this long list of dos and don'ts, but as the story of how a loving God keeps his promise.

I read over and over about a God committed to loving me. How James, the brother of Jesus, in chapter two verse thirteen, showed me that mercy triumphs over judgment. I read how Paul explained to the church of Rome that it was the goodness of God that leads people to repentance: "Or do you despise the riches of His goodness, forbearance, and longsuffering, not knowing that the goodness of God leads you to repentance?" (Romans 2:4 NKJV). Why weren't more Christians talking about this? Why were we trying to make the judgment of God bring about repentance? Why were we so concerned with crafting messages on the wrath of God that would swallow us whole if we didn't get things together? Why weren't we talking more about the goodness of God? The Bible freed me. I understood why the early church fathers chanted "the Bible and the Bible alone."

The book *How God Changes Your Brain* says that what I experienced is what research predicts will happen. "Our research suggests that the more a person contemplates his or her values and beliefs, the more they are apt to change."[6] I was shocked. I thought the sign of a faithful Christian was my being concrete on my understanding. Apparently, it is the opposite. The more we are willing to wrestle with the God of the Bible, the more our understanding will morph. Conversely, people who are unwilling to give any consideration to the "what if" questions in Christianity are the ones who have spent the least amount of mental energy trying to wrestle with what they believe. The more I read Scripture for myself, the more I started to see how other people could read it differently. I became less judgmental of other denominations.

This fits with what theologian James Fowler describes as the "stages of faith development." He lists six stages. Stage 0—Primal Undifferentiated faith—is the stage where we experience trust and mistrust of religion. Stage 1—Intuitive-Projective faith—is the stage where we acquire language for our faith, which leads to Stage 2—Mythic-Literal faith, the stage where we develop ideas of justice and fairness. If certain promises we see in Scripture don't come to pass, it may make us feel like God is angry and cheating us if we are members of Stage 2.

Stage 3—Synthetic-Conventional faith—is where if someone starts to challenge our belief system, we see it as a threat. I think I lived in Stage 3 for years, struggling with anyone who didn't agree with me. Trying to force people to land on my same convictions. Questioning the salvation of others who didn't interpret complicated biblical texts the same way I did. These are the "but the Bible is so clear on this issue" people. If you don't know one, it's probably you.

Stage 4—Individuative-Reflective faith—which is where religion starts to take on greater flexibility and nuance. This is what happened to me the more I read the Bible. Rather than become more rigid, I became more flexible. I started to see people as human beings who had been trained by different thinkers and experiences than me. Of course, we may arrive at different destinations—we didn't have the same journey. Stage 5—Conjunctive faith—is where Fowler says a participant may move beyond traditions to hold a multidimensional view of who God may be, and truth as something that can't always be articulated (shout "amen" if we are tracking now). Lastly, Stage 6—Universalizing faith. This stage is rarely reached. It is where one sees humans as transcending religious differences and beliefs, but sees all human beings regardless of differences as worthy of compassion and understanding.[7]

It's crazy because you would think the more you read the Bible, the stricter you would become, but it was the opposite for me. This is also because "the majority of spiritual practices suppress the brain's ability to react with anger or fear."[8]

So why do we have so many committed Christians on Twitter telling us we are all going to hell? The fruit of the Spirit is not metaphorical, it's backed by science. The more time we spend reading God's Word and meditating on his promises, the less likely we are to feel threatened when someone disagrees with our faith. Paul writes in Galatians 5:22-23, "But the fruit of the Spirit is love, joy, peace, forbearance, kindness, goodness, faithfulness, gentleness and self-control." This isn't spiritual crazy talk; this is neuroscience. And when Jesus says in John 13:35, "By this everyone will know that you are my disciples, if you love one another," he isn't being romantic; he is being God.

And you know how it says in Romans 12:2, "Do not conform to the pattern of this world, but be transformed by the renewing of your mind"? This also is science. Newberg and Waldman explain that thinking about God changes our brain physically. Just like choosing to focus on learning how to play an instrument changes our neural circuitry, time meditating on the Word of God enhances our cognitive health. "But religious contemplation changes your brain in a profoundly different way because it strengthens a unique neural circuit that specifically enhances social awareness and empathy while subduing destructive feelings and emotions."[9] And then two neuroscientists write, "This is precisely the kind of neural change we need to make if we want to solve the conflicts that currently afflict our world."[10]

Here is another reason we should all freak out that 72 percent of Christians serve an angry God. Apparently, the God you believe in is the God you become. According to a Baylor study, when people saw God as authoritarian, they were more likely to believe in the death penalty.[11] They believed the government should have a lot of power over its citizens and were quick to dump money into funding for the military. Authoritarians tend to respect authoritarian-type leaders and are more likely to follow orders for the sake of following them. They condemn people who challenge the status quo. They are less forgiving and tend to respond with vengeance toward sinners, and they are also more likely to raise children who are unwilling to accept anyone not in the in-group.

The hard part is that if you are authoritarian, you may not even know it.[12]

> If you tell people about authoritarianism, including the part about authoritarians being aggressive when backed by authority, and then ask them how willing they would be to help the federal government eliminate authoritarians, then—you guessed it—High [authoritarians] will be more willing to volunteer than others, to hunt themselves down . . . And yet, compared with most people, they think their minds are models of rationality and self-understanding.[13]

While religion has been a source for much good in our world and society, in the wrong hands and worshiping an angry God, it can do a lot of damage. The same religions that gave us Martin Luther King Jr., Nelson Mandela, and Mother Teresa have also produced extremist groups and alt-right religious groups based on White supremacy. I saw someone tweet the other day, "Honestly the word Christian is becoming a red flag for me." But it isn't religion that is bad, it's the promotion of an angry God. Our enemy is not religion, it's the spiritualization of anger, exclusion, and extremism.

The good news is, this is not the gospel. In Matthew 22:36-39, a lawyer asks Jesus, "Which is the greatest commandment?" and Jesus answers, "'Love the Lord your God with all your heart and with all your soul and with all your mind.' This is the first and greatest commandment. And the second is like it: 'Love your neighbor as yourself.'" The lawyer then asks, "And who is my neighbor?" (Luke 10:29) and Jesus tells the story of the Good Samaritan. The Jews hated the Samaritans. You are talking about hundreds of years of fighting, disunity, and racial prejudice. Basically, Jesus said, "Your neighbor is whoever you hate the most." Love summarizes the entire gospel: love toward God and love toward people. The Ten Commandments can be summarized as the laws of love: the first four commandments, love to God; and the last six, love to others. I would go as far as to say your relationship to God is only as deep as your relationship to people. Love wins.

This is great news and is shared by the 23 percent of Christians who see God as forgiving and gentle. They are more likely to activate areas of the brain allowing them to become more loving, gentle, and forgiving. When we focus on an angry God, the parts of our brain that respond with fear and anger are activated. But a loving God activates our anterior cingulate, which actually suppresses our brain's response to fear and anger. It even aids us in creating emotions like empathy toward those who are hurt. Anything we focus on sends signals to our brains that cause a response. This is why we have to re-envision God.

I didn't just need a God that was loving in order to feel loved, I needed to worship a God of love in order to give it.

When I learned this, it became incredibly important for me to start educating Christians on the true character of God. If love is who God is, then us knowing love is directly correlated to us knowing God. The antichrist would also be anti-love. This would be the complete opposite of God's government and how he chooses to rule. Love is not a small, take-it-or-leave-it Christian buffet choice, it is literally who God is. It goes against the very nature of God to not love you. God cannot help but be totally in love with you because God is love. It's imperative that we free churches from the distant, critical, angry God. These churches are producing distant, angry Christians. John 13:35 says, "By this everyone will know that you are my disciples, if you love one another." Peter Scholtes wrote a hymn in the 1960s based on John 13:35 declaring, "And they'll know we are Christians by our love, by our love. Yes, they'll know we are Christians by our love."

God is love. And if I err and am doctrinally wrong on an issue, I would rather err on the side of love.

Research reveals if we take twelve minutes a day to focus on a God of peace, love, or mercy, in just a couple months, our brains will be rewired toward compassion and even less likely to reject people who believe differently than us. Twenty minutes to an hour a day of this type of meditation rewires our brains. I meditate on scriptural promises or character-istics I know God has. When we do this, we are neurologically changed. As love goes up in our brains, fear goes down. I used to serve God out of fear, but now I serve God out of love. And I finally understand why there is freedom in Christ. It may be time to ask yourself, which God are you serving? Do you see

God clearly? Do you need to re-envision God? Or do you see God as angry? This matters because it impacts the type of follower of God you actually are.

Re-envisioning God

On Memorial Day of 2020, my husband and I had a fight, and then he went hiking. Nature is how he gets perspective. Walks and hikes together are a fairly big part of our relationship. It's not that I love being outdoors, it's that I love him, and he loves being outdoors. On May 25, Seth hit a trail with our seven-year-old son, Hudson. He was overwhelmed by pastoring during COVID. He felt like God led him to this church and within six months, it seemed everything was falling apart. The entire staff quit but him, and he had to go from being an associate youth pastor to interim head pastor during a pandemic. And by the way, this was his first year as a full-time pastor at all. Our tensions were high; I was frustrated, and he needed air.

A few hours later, he called me and apologized for the fight. He sent some pictures of himself and my son, and we fell madly in love again.

A few minutes later, my phone rang. I was surprised he called me back so quickly.

"Hi, Heather. This is Ryan, and I'm a first responder." My breathing stopped. "There has been an accident." Everything froze.

Ryan went on to tell me that while walking the trail with his fiancé, he heard a scream for help. My husband had been standing on a boulder that came out from under his feet. He rolled about thirty feet down a hill, and then fell ten feet off a cliff.

When I got to the hospital, the doctor told me that best case scenario, Seth had a broken pelvis. They ran scans and didn't find anything wrong, so they sent us to a bigger hospital to do more tests with better equipment and imaging. While we were sitting and waiting for his results, I imagined my life with him paralyzed. I also imagined my life if it were a crushed vertebra and he would have to be bedridden for several months. I kept praying for strength because I knew how hard this was going to be. We have no family in Denver and a church he had just come to.

"No matter what happens," I said to him as he lay there under crippling anxiety that he would have a spinal injury, "we leave here together. Whether you can't walk ever again, or for six months, we will do it together."

The doctor came in. "You are extremely lucky," he said after running another test. "There doesn't appear to be a single crushed vertebra." He was shocked. My husband fell ten feet off a cliff and slept in our bed that night. But if he hadn't, no matter what ways his life would have changed, we would learn to change together. I am not married to my husband because of what he can do for me. I am married to him because of who he is. Our marriage isn't built on money (we have none) or cars or stuff. The best part of our marriage *is the marriage*, it's the relationship, it's knowing we love each other.

The best part of God is not whether or not he blesses me, it's God. It's knowing that feast or famine, we get to set the table for two. Rather than the blessings being determined by God's presence, I now believe God's presence is the blessing. The joy of the Christian is bigger than any singular answer to prayer, it's to live in a perpetual relationship with God himself. The circumstances of life may bring incredible disappointment, but he is the rock we get to build our lives on.

"But no matter what happens," God says to each of us as we sit in our waiting rooms, "we leave here together."

> ### *Promise ten to memorize:*
> "And we know that in all things God works for the good of those who love him, who have been called according to his purpose." (Romans 8:28)

DISCUSSION QUESTIONS

- ► How do you view God? What are God's characteristics?

- ► How do you know God is this way from personal experience? Where in your life have you seen these characteristics displayed?

- ► How much time do you spend daily with God? Are you happy with that time frame? Should it be more?

11

It's Your Turn to Move on Maybe

At some point I just decided that there would be no guarantee outside of maybe. Maybe this could work. Maybe this could be something. Maybe I'm not crazy. Sometimes you just have to move on maybe.

HEATHER THOMPSON DAY

WHEN MY HUSBAND AND I had been dating for three months, I decided I couldn't take it anymore. We were sitting in a Subway, a romantic date only college kids can afford, and I looked him in the eyes and said, "I love you." I can still see his face in my mind, beaming. He had been telling me for weeks, "Heather, I really care about you." At the time I thought this meant he really liked me, but he didn't love me. I didn't care either way. I loved him, and I needed to tell him that. Now that I know my husband better, I realize "I care for you" was the equivalent of him shouting his love for me from the rooftops. He is introverted and resists vulnerability at most turns. Honestly, I am not sure if Seth really knew how to say I love you until he was certain it would be reciprocated.

I've always been the kind of person willing to say I love you first. I've never known how to *not* put myself out there. This has certainly burned me at times. I mentioned earlier I have been rejected more than I can count. I have had my heart broken. I've told people something was coming and then felt humiliated when everything fell through, but this is still the only way I know how to live. I am willing to trade comfort for possibility, and vulnerability for opportunity. For me, my biggest fear has always been failure. Because of this, I am willing to try things that may burn me if the reward will be perceived success.

For example, I emailed civil rights speaker and antiracist Jane Elliott, creator of the "blue eyes, brown eyes" experiment. She has been featured on Oprah, and if you don't know who she is, please go to YouTube immediately;[1] the video has been viewed over one million times. In fact, there is even a movie about her regarding how she divided her classroom between brown-eyed and blue-eyed kids the day after Martin Luther King Jr.'s assassination. Within days, perfectly well-scoring blue-eyed kids started to perform worse than their brown-eyed counterparts. The only reason was because their teacher, Jane Elliott, told them brown-eyed kids were smarter. She was making a point about racism, and all of America watched.

I have zero business emailing Jane Elliott, but I did. I asked her if she would be willing to hop on a Zoom call and give a lecture to my intercultural communication class that semester. I had no idea what she would say, but I willingly put myself out there because I respected her work and thought it would be an incredible opportunity for my students. One night my phone rang and, to my shock, it was Jane. She wanted to book the lecture. She spoke to my class every semester after that for a couple years and even provided me with an article for my personal blog.

When people asked, "How in the world did you get Jane Elliott to lecture for your class?" my answer was always the same; I asked her. I was willing to move on a maybe.

This last time I went on my yearly Scripture read, I came to John 1:35-39:

> The next day John was there again with two of his disciples. When he saw Jesus passing by, he said, "Look, the Lamb of God!"
>
> When the two disciples heard him say this, they followed Jesus. Turning around, Jesus saw them following and asked, "What do you want?"
>
> They said, "Rabbi" (which means "Teacher"), "where are you staying?"
>
> "Come," he replied, "and you will see."

I nearly fell out of my chair. This Jesus of Nazareth was the same God I experienced in Denver. A God who didn't give me a road map, didn't provide me with the full picture, and felt vague in telling me where I was going. When Jesus tells the disciples, "Come, and you will see," I know that he is the same God yesterday, today, and tomorrow. I recognize those words because God rarely reveals our ending at our beginning.

Stephen King's first big novel, *Carrie*, was rejected thirty times before he finally got a book deal. I remember reading at one point his wife fished it out of the trash. He didn't know it would be a massively successful book and movie. There is simply no map for maybe. John Grisham's first novel, *A Time to Kill*, was rejected twelve times. How hopeful do you think he felt after rejection number eight? Do you think J. K. Rowling, who was also rejected twelve times, thought she was on the brink of becoming the most successful female author in all of the United Kingdom with her books selling nearly a half billion copies? I recognize moving forward without all the pieces in place may be easier for some than others. Some people hate risk. Even the possibility of rejection is enough to make you quit. I want you to push past that. You may be uncertain, but I want you to do it anyway. At some point, I just decided there would be no guarantee outside of maybe. *Maybe this could work. Maybe this could be something. Maybe I'm not crazy.* Sometimes you just have to move on maybe.

While you may feel out of control and reckless, some studies show risk takers are quite the opposite. A study at the University of Turku in Finland wanted to understand what happens neurologically during decision making. They looked at the brains of thirty-four young men ages eighteen to nineteen, and divided them into two groups: high-risk takers and low-risk takers. They hypothesized people who spent a lot of time assessing the risks of any given situation would have more highly developed neural networks. The brains of the high-risk takers were different from the brains of low-risk takers in the areas called "white matter." People who took chances and made quick decisions during a driving simulation had significantly more white matter than those who took more time. It "took brains" to take risks.[2] Interestingly, WebMD writer and editor Jennifer Warner writes, "People who enjoy taking risks may be more content and satisfied with their lives."[3]

In a Cornell research article, editor Jackie Swift talks about the risk-taking behaviors of adolescents. The typical mode of thought about young people is they are stupid. They take risks because they don't correctly weigh the pros and cons. Swift doesn't believe the risk-taking behaviors of adolescents are impulsive at all. Young people are willing to take calculated risks. If it seems the odds are in their favor, they are willing to take the leap. Adults, however, seem to lose this.[4] Even when the deck is stacked in their favor, they are less likely to engage in risk. An article on Entrepreneur.com says being able to achieve success is almost always linked to risk taking. They quote Mark Zuckerberg, the founder of Facebook, as having said, "The only strategy that is guaranteed to fail is not taking risks."[5]

Being Brave

New York Times bestselling author Brené Brown says, "The key to whole-hearted living is vulnerability. You measure courage by how vulnerable you are."[6] Brown made the topic of vulnerability famous after her TED Talk went viral. Her research found that while vulnerability is the birthplace for pain and rejection, it is also the building block for joy, love, and belonging. You can't experience love if you aren't willing to experience heartbreak. You can't experience success if you aren't willing to risk failure. Being brave is about being willing. While our brains are quick to tally the costs of our action, they rarely process the costs of inaction. How often do we ruminate over the trips we didn't take or the conversations we didn't start? We worry what will happen if we ask, but the reality is, if we never ask, the answer is always no.

Brown describes vulnerability as the willingness to show up even when you don't know the outcome. I am calling it the ability to move on maybe. One of the greatest takeaways I experienced from Brown's work is this concept she calls "whole-hearted" and how some people live bravely, being willing to step forward with their whole hearts. When trying to discover why some people were able to experience love and belonging and others seemed to constantly be searching for it, Brown found only one distinguishing factor: the people who had it also believed they were worthy of it. What if you are worthy of it? What if this perceived risk is actually an opportunity? What if you moved on maybe?

Being brave is a topic Brown talks about a lot. She describes her daughter joining a swim team. Her coach put her daughter in the 100-meter breaststroke, and her daughter was extremely nervous because it's a hard race and also not her strong suit. Her daughter wanted to quit. There was no way she would ever win the race. Brown told her, "You will never win this race, but maybe winning for you is getting off the block and getting wet." When the race came, she did get wet. Brown said the meet was brutal, and her daughter came in so far behind that the swimmers were waiting for her to finish so they could start the next race. When she reached her mom she said, "But I was brave, and I won."[7]

That story has sat on me like a down comforter. What if trying, even if we lose, is what actually causes us to win? What if your turn is not even about the destination as much as you going on the journey? Every time we are willing to get back in the water, we risk the possibility of drowning. But what if all that separates winners from losers is getting off the block? What if being brave is actually the way we win?

Talking to Strangers

In the book *Talk to Strangers: How Everyday Random Encounters Can Expand Your Business, Career, Income, and Life*, David Topus talks about the discovery he made just by talking to people sitting next to him on a plane. He started chatting with whomever was seated next to him and discovered many of these connections became priceless. He'd open just with basic conversation, and then share with them what he did professionally. Often times, he would find the person sitting right next to him on a random flight was a potential client. He made some of his strongest business deals on airplanes, and says it made him realize how much we miss by not engaging the people around us.[8]

He eventually started booking flights from city to city without any actual reason to be there. He just knew these flights would bring him a couple hours of the attention of his seatmate, and for Topus, these flights often produced his next business deal. He believes every time you meet someone, you should assume they may hold the key to what you need either personally or professionally. Often we miss incredible opportunities to network because we aren't willing to be brave enough, or vulnerable enough, to simply talk to the person next to us. While most of us don't have the financial resources to book flights with nowhere to go (Hi, I am most people), the concept is fascinating. What if we took advantage of everyday encounters? What could we have discovered if we had only put down our phones?

Of course, you don't want to force yourself on someone who clearly doesn't want to communicate, but if you are in a long line, and the person in front of you isn't preoccupied, why not force yourself to try to have a conversation? At the very least, you are training your brain that you are willing to take risks. What if when you are sitting at the park watching your kids play, you struck up a conversation with whomever else is there? What if we put our phones down, and noticed who was around us? These small choices, made over and over, could eventually lead you to big reward. Sometimes, you have to move on maybe.

The Five-Second Rule

Mel Robbins, author of *The 5 Second Rule* and the TED Talk *How to Stop Screwing Yourself Over*, writes and speaks about the major premise of her rule being if your brain gives you an idea or instinct, you have approximately five seconds to do something about it physically before your brain kills it. Everyone has a window between their brain giving them something they want and then coming up with reasons as to why it won't work, she says, and that window is only five seconds long. Robbins says as soon as inspiration strikes, you have to follow it up with something tangible, even if it's as small as writing it down into your phone. Our brains hate change, so knowing you have five seconds to make a change is a really powerful incentive to get up and move.

I try to implement this concept into my everyday life. When I think about how I'd like to have lunch with a friend, I send the text to get the ball rolling rather than run through all the work obligations I have that would eventually have me not reach out. When I have a great idea for a sermon or lecture, I immediately write it into the Notes app on my phone so I have created documentation for it. Taylor Swift says when she gets inspired by the hook for a song, she stops what she is doing and puts it into her phone and has found herself at 2 a.m. singing a hook into a voice memo.[9] When you get inspired, Robbins says we must act in some way toward that inspiration. If someone new shows up at church, walk over right then and introduce yourself. You have five seconds to put some legs on whatever idea you are debating.

Robbins goes on to say:

An instinct is not buying everybody in the bar a round of shots. An instinct is not a rash, irreversible decision. It's not destructive, illegal, or harmful behavior. I define an instinct as any urge, impulse, pull, or knowing that you should or should not do something because you can feel it in your heart and your gut."[10]

Let's say right now you have an instinct to send a follow-up email after a conversation that made you excited about a possible partnership today. You have five seconds to do something about it before all the excuses kick in as to why it won't work. Robbins calls your gut your second brain. While some people may mock those gut reactions, she says we should lean into them. In fact, we have to often push ourselves toward them because our brain likes to keep things the same. It doesn't like to break routine. If you want to get healthier, your gut may say, it's time to go for a run, but if you don't grab your shoes within five seconds, you aren't going anywhere. So, what is your gut telling you right now?

5. . . . .

4. . . .

3. . .

2. .

1.

Move on maybe.

Praying on Maybe

Maybe you aren't ready to jump in. Are you at least ready to pray? Maybe with your five seconds, prayer can be the first action step you take. Should you quit this job? Make an offer on that house? Start that writing project? Enter that graduate program? Ask them on a date? Maybe you are ready to surrender this decision to God and ask for his involvement. I believe prayer matters. I often hear Christians say things like,

"I don't think prayer changes anything," and I respectfully believe they are wrong. Words are powerful, even outside of prayer. Within it, they can be miraculous. Can you commit to at least praying on your maybe?

Some of the research on the mind and how it influences objects is, pardon me, mind-blowing. Let's take for example the work of Dean Radin. Radin believes human consciousness can transform the physical limitations of the body and mind. In other words, Radin believes, from a research standpoint, our thoughts hold power.[11]

Radin did a double-blind study at the Institute of Noetic Sciences on the effects of human intention on another person's autonomic nervous system. The study found that someone sending intentional, compassionate thoughts produced changes in the skin conductance of cancer patients, even when they were nowhere near the thought sender and were totally unaware of the experiment. Radin calls it compassionate intention; Christians call it prayer.[12]

While I believe in the power of prayer as a woman of faith, I also feel encouraged by it as a person of the academy. That said, the elephant in the room is the idea that prayer works best when followed up by our own action. God will do what you cannot, as long as you do what you can. We can't just pray for the people of Flint to get clean water. Christians should pray *and* move. God wants to colabor with us in the gospel. That is the entire point of Christianity. We are to organize and mobilize to alleviate human suffering.

Christians should not use prayer as an excuse to stay seated. "Thoughts and prayers," without movement and action is not exactly colaboring. We can't keep watching tragedies and sending our thoughts and prayers when we know our mobilization would be more effective. It reminds me of a verse in James 2:19 that says, "You believe that there is one God. Good! Even the demons believe that—and shudder."

At some point, church, our belief is not enough. And our prayers, without our actions, can only go so far. Or, if I can put it into Christian lingo, our faith, without works, is dead. I do get offended when I see people online mock politicians or religious leaders for their prayers. I believe in prayer. I have experienced its bounty. But I also think Christians should try to understand what the secular world has watched for years: a church that says one thing and does another. A church ridden with sexual abuse scandals, misogynistic treatment of women, and downright hostility toward the LGBTQ community. It's hard to watch people who say they believe in peace and love be so unloving and wreak so much havoc. The world needs Christians who have had real experiences with Christ. Because a true encounter with God changes you. Experiences can't be studied. Experiences are not data sets. And yet they change everything. If you do nothing else with whatever your next move is, can you at least pray about it? Pray for God to open what should be opened, and close what should be closed, and stir your heart if it should be stirred.

It's Your Turn

The psalmist writes in Psalm 139:13-16:

> You created my inmost being;
> you knit me together in my mother's womb.

I praise you because I am fearfully and wonderfully made;
   your works are wonderful,
   I know that full well.
My frame was not hidden from you
   when I was made in the secret place,
   when I was woven together in the depths of the earth.
Your eyes saw my unformed body;
   all the days ordained for me were written in your book
   before one of them came to be.

Listen (I'm clapping now), God cannot plan your life if he hasn't also planned your days. And this moment matters. The devil wants you to believe this space is insignificant. He wants you to give up and go home. He doesn't want you to discover that the waiting room is holy ground. God is perfecting his perfect work in you that you may be complete, lacking nothing.

If you feel a stirring in your soul right now, I want you to get up and take one action step forward toward the purpose you feel called to. Maybe that's sending an email, maybe that's getting on your knees and saying *okay God send me*, maybe that's setting a goal to wake up twenty minutes earlier and have worship tomorrow. I don't know what that step looks like for you, and maybe you feel horrified and paralyzed and unsure of how you can pull this thing off. Do it anyway. Move on maybe.

Promise eleven to memorize:

"God is not a man, that he should lie; neither the son of man, that he should repent: hath he said, and shall he not do it? or hath he spoken, and shall he not make it good?" (Numbers 23:19 KJV)

DISCUSSION QUESTIONS

► What does it mean for you right now to start living bravely?

► What in your life has God asked you to "come and see?"

► What might it be your turn to do?

12

It's Your Turn to
Make Your Move

And the day came when the risk to remain tight in a bud
was more painful than the risk it took to blossom.

ANAIS NIN

I N HIGH SCHOOL, I had a nickname: Oprah. They weren't
saying it to be affirming. People were mocking my goals. I
wanted to be a journalist. I was an editor for the school paper
and my friends could not understand for the life of them why I
pulled students out of class to get quotes for a good story.

"She thinks she's Oprah Winfrey," a boy said in history class,
and everyone erupted in laughter.

"Hey, Oprah!" It stuck. The unfunny thing is, I did want to
be Oprah Winfrey. She was Black, she was successful, and she

was a storyteller—it was all I could ever dream of. I watched the Oprah Winfrey show every day after school. I loved how she inspired us all to value conversation. I loved how she built bridges with so many different racial groups. I loved how she talked about things people weren't talking about, but always did it in a way that left you wanting to be better. But then I would go to school and hear, "Hey, Oprah!" shouted across the cafeteria. Every time they said it, I could feel the weight of my dreams sink beneath their laughter. I finished a bachelor's in journalism, but I never did become a journalist. I still dreamed to be like Oprah though. I wanted to tell stories that made an impact in our culture. I wanted to be an example to kids that looked like me, that you could grow up in a small town with two stoplights and still dream big dreams and create room at the table for other marginalized voices when you got to where you were going.

A couple years ago, I was scheduled to do a breakout session for a teacher's convention in Chicago. Thousands of people were there, and I decided to go to the main session in the giant auditorium because I heard gospel singer Wintley Phipps speaking and singing. I took my seat and, as his full voice filled the auditorium, I wept. He reminded me of my dad who was a singer, and whose giant voice had filled so many churches and my childhood. Now, my dad's health wasn't great, and I missed hearing him sing. I wept because I just felt like I wasn't where I wanted to be in my life. I worked full-time as an assistant professor, paying back student loans that were triple my annual salary. I had always dreamed of being a writer of books that made a difference in culture, and though I had published six books with a small publisher, I don't think any of them ever sold more than a few thousand copies. I felt defeated and overlooked in so many areas of my life, and as I heard Wintley sing, all of my emotions and everything I had been putting a positive spin on for months felt like it was crashing down all around me. I had to speak in thirty minutes, and here I sat sobbing to Wintley Phipps in the back of the dark auditorium.

"I was singing in Baltimore at the Civic Center," Wintley began speaking slowly in between songs. "I came off the platform and felt a tap on my shoulder." He went on to say he saw a woman standing there asking if she could speak with him. "I need to talk to you," she said. The woman was a frustrated reporter. She made $25,000 a year and was overworked. Overwhelmed by the concert Wintley just gave, she felt it wasn't her turn. She needed encouragement and prayer. Wintley prayed with her, and at the end he said, "Before you go, God has impressed me to tell you he is going to bless you, and give you the opportunity to speak to millions of people."

The woman looked at him and said, "Do you really think God would do that for me?" It seemed inconceivable. Years later, that woman would become nationally recognized—as Oprah Winfrey. She has helped fund Wintley's nonprofit work helping kids in inner cities, and the two have stayed connected both personally and professionally, all because Wintley agreed to stop for a twenty-five-year-old woman frustrated with her life at a concert and looking for prayer.[1]

The story was so powerful to me. It had never occurred to me that one of my heroes, who in my eyes was so brilliant and successful, would have felt dejected and defeated. I couldn't believe there was a point in Oprah Winfrey's life where she lived off of someone else's prayers for her. It dawned on me that even the Oprahs have had days when it wasn't their turn.

I think stories are incredibly powerful. They bring connection into loneliness, hope into fear, and intimacy into isolation. When you get to where you are going, don't forget to tell the story. None of us are ever truly alone in our experiences. Others have wept like we wept, been crushed as we were crushed, and hoped as we hoped. What if life isn't about reaching a particular destination as much as it is about living the story? Here are some examples of people who are living theirs.

The Power of the Storyteller

Scarlett Longstreet

When I was four my parents placed my newborn sister for adoption. When my parents first told me, I kicked the dashboard of our car until my legs bruised. A tiny protest. I stared out the window into the woods while my mother labored, wondering how you make grownups change their minds. I've felt her absence my entire life.

After years of almost dropping letters in the mail, typing and then deleting Facebook messages, and staring at the same few pictures several hundreds of times ... I put a note on her doorstep. I told her I hoped to know her. I told her about myself. I worried every word I wrote was wrong. She texted me the next day. But she wasn't ready to meet me. I've longed my entire life to love that little girl, but it's still not my turn. But I'd rather sit with rejection, I'd rather sit in the waiting, than sit in the what if.[2]

Seth Day

I failed at college several times before I would be a successful student. The thought *you will never be good enough* was just part of who I was at the time. It didn't matter how hard I tried; I couldn't stay focused long enough to make the grades needed to pass my classes. What contributed to my lack of focus was my internal grief over the loss of my older brother who passed away less than a year prior.

Sometimes, it's okay that your turn hasn't come yet. It's okay that you haven't gotten your diploma, or a fancy job promotion. It's okay that the world seems to move forward while you're left feeling stuck in the mud. Why? Because it may be your turn to mourn the people, and things that once were, but are no longer a part of your life. Maybe it's your turn to grieve, or breathe, or cry. It may not be your turn for success by the American standard, but maybe it's your turn for healing.

Because when your turn comes, you want to meet it with the best version of you that exists.[3]

Vimbo Zvandasara

Up until recently, I had always imagined that my autobiography would be titled "The Life of a Third, Fifth, and

Seventh Wheel." This existence of being single and tagging along with your "couple friends" is obviously not a coveted relationship status. To be single in a world full of epic-viral wedding proposals, social media saturated engagement and wedding photos and notifications can be tough at best, and downright depressing at worst. I have seen a lot of what dating in the modern world has to offer. I fit the twenty-seven-bridesmaids-dresses-and-tried-every-dating-app-that's-kosher stereotype. I have also spent a lot of time wrestling with the thought, *When will it be my turn to love and be loved, God?*

Recently, I've been surprised by love. And right in the midst of being miserable about being single and also praying to God to just let me know if I was going to be forever single, I have now stumbled into a relationship. And he is great! He laughs at my jokes, we share the most exciting and boring parts of our day, and quite honestly, he really loves me well. I know that's probably weird to say, but similar to a friendship that just fits, dating the right person makes life oh so sweet. It wasn't my turn for *years*. But right now, it feels like my turn is arriving, and suddenly it doesn't matter so much that it hadn't been my turn all that time (I never thought I'd say that). I'm just so consumed in what I have found now.[4]

Carla Hernandez

I was born in the United States, but I am a first-generation student. My parents emigrated from Mexico. In my house we only spoke Spanish, and so English has always been my second language. All throughout elementary school I could understand English, but I couldn't really speak it. I remember being in sixth grade and just responding "yes" and "no" and trying to keep any teacher-student interactions brief so people didn't know I couldn't actually speak English even though I could read it. My dream was to be a nurse. My parents had given up so much to be in this country and worked hard jobs so that we could eat. Being a nurse in my mind would show them that their sacrifices were worth it.

I finished nursing school. It was a struggle for me, but I finished. When it came time to take the NCLEX, I was nervous. I've never done well with timed tests. It created so much anxiety. My parents had helped pay for me to go through college, and this test was the last thing standing between me and everything they had done to get me to this point. I didn't get my results for months, and when I finally got them, I had failed. My first reaction was, "Okay, I just need to focus more. People fail. It happens." I took it again, a whole year later, and at this point I'm starting to feel like maybe I'm not supposed to be a nurse. Maybe this is God telling me I need to go in a different direction? I failed again. I wanted to be done. But my parents encouraged me to do it again. My husband had me quit my job so I could just focus on studying, and a woman from church met with me once a week to practice.

On my third time, after so much prayer and perseverance, I passed. I can't even describe what relief swept over me. I got a job within one week of passing and today, I am the first person in my family to graduate college.[5]

Alex Gordon

The famous Notre Dame ambassador Lou Holtz said it best when he wrote, "You don't go to Notre Dame to learn something, you go to Notre Dame to be somebody." I wanted to be somebody. My application to Notre Dame's law school showed that I was a thirty-two-year-old nontraditional applicant with an impressive resume and an above-average LSAT score. What it failed to communicate is that I was married with two children under the age of three and an eleven-year-old stepson. I chose Notre Dame because it was not only one of the top law programs but also because it would allow me to not displace my stepson from his biological father. It was close to where we all lived, and I wanted as much as possible to give him continuity.

As a stepfather, my relationship with my stepson is ever-evolving. It is a balancing act and I am learning on the fly. Unlike other situations, I must react to his feedback and consider how to convey certain messages. Already at age eleven he has begun to question his identity: Who am I? Where is my place? These are delicate questions and I do not presume to have all the answers. You see, he is Black and I am half Asian and half White. There are important roles I can play and lessons I can teach, but there are others best reserved for his father. Despite the issues his father and my wife have, he is a good father and a strong role model to our son. He actively seeks to remain an integral part of his life, which is critical as my stepson seeks to find himself. I thought surely Notre Dame would be the school I got into, until I didn't.

The devastation I felt getting the letter that my application had not been advanced is hard to describe. I was trying to go back to school, trying to better my family, trying to make it as easy as possible for my stepson, and the pieces just didn't fit. I was balancing so many different plates and the one I wanted the most crashed. The good news is that I still went to law school. It wasn't my dream school but it was a good school. Then, I got offered a three-fourths tuition scholarship. The best part was the program allowed me to drive to Chicago every other weekend so that my stepson could stay in the same place. One door may have closed, but I am thankful for the one that opened.[6]

Flor Osorio Just

I had no clue what endometriosis was. My mom never told me about it, nor was it ever discussed among my girlfriends. I didn't even learn about it in pharmacy school. On a gloomy October day, when my OBGYN told me I had stage IV endometriosis, I was stunned.

The doctor said I needed to have surgery and that my severe endometriosis put me at a higher risk of developing cancer. A hysterectomy was likely in my future, and most importantly, I would have trouble having children of my own naturally. Suddenly, my throat got extremely dry, my eyes a little blurry, and my ears began to ring. I cried tears from the depth of my soul that day. My world was never the same after that.

As I would later find out, my best friend was dealing with something different yet completely related. Let me explain. At the same hour I was sitting in a doctor's office coming to terms with my incurable disease and my infertile future, my beloved friend found out she was unexpectedly pregnant. And just like that, my perfect world was turned upside down. The diagnosis sent me into a whirlwind of anxiety, depression, and grief. It affected the way I viewed my self-worth, my body image, my friendships, and even my marriage. I was angry. I was mad at God for allowing this to happen to me. But I was infuriated at myself for letting it take control of my life. In the midst of all the sadness and confusion, I lost my balance. I made choices that I thought would fill the void in my heart, but instead all it led to was the loss of my closest friends and my marriage on the brink of divorce. I needed God desperately.

After going to Burning Man to "find myself," I booked a solo ticket to see my family in Costa Rica to decompress. It was there that my mother forced me to see a specialist. After reviewing my chart, the doctor said to me, "Your endometriosis is severe and you need to have surgery. But that doesn't mean you can't have children. Do you believe in God?" He pulled out a Bible from his desk drawer. "When I first started practicing medicine, I would tell women who had endometriosis that they couldn't have kids. And yet, somehow, they still got pregnant. Don't let anyone tell you what only God can."

That very next day I found myself in a Costa Rican OR recovery room with a huge weight lifted off of my uterus. Literally! God had begun to refine me, from the inside out. He opened his arms of forgiveness and grace and engulfed me in it. He opened my eyes to a new life and restored my marriage in the process. He saved me.

Today, I embrace the mess that God turned into a masterpiece. What the enemy meant for harm, God has used for good. Although it may not be my turn to be a mother yet, I know God has a beautiful plan for me.[7]

Mary Beth Minnis

Ephesians 2:10 says that we were "created in Christ Jesus to do good works, which God prepared in advance for us to do." It comes immediately after "For it is by grace you have been saved, through faith . . . it is the gift of God—not by works, so that no one can boast." In each season of life, God has led me to walk out the good works he has prepared for me, and usually it means I need to wear a different shoe.

Since 2010, I have played a role in producing several documentary films. I love the privilege of telling stories of hope in seemingly hopeless situations. And in this particular season, I am often wearing sparkly shoes on a red carpet at a film festival or award ceremony.

When I joined the *Jump Shot: The Kenny Sailors Story* film team, it was a documentary in the making about an unknown basketball player. I began praying that God would give us favor with NBA stars. I went to the mall and tried on LeBron's shoes as well as several others. LeBron's made me feel like a beast, like a lion, and like a king. My favorite pair were Kevin Durant's. I felt agile, ready to jump, and they just seemed to fit perfectly.

I started walking while wearing my red KDs, and praying, asking God to give us favor with Kevin Durant. Six months later, one of KD's mentors, myself, and two other *Jump Shot* producers were invited to show the film to him. It was incredible. I was struck by Kevin's humility, his teachability, and his heart for Kenny Sailors. Kevin is featured in *Jump Shot* and I think you get a glimpse of his heart that the media often misses in their portrayal of him.

What shoes are required for the season of life you are in right now? Do you have a favorite shoe that gives you confidence to walk out his good works for you?[8]

Your Turn Starts Now

Maybe it's not your turn today. Maybe you've been overlooked and underappreciated. Maybe you are tired of seeing wedding invites on your calendar but no ring on your own finger. Maybe it feels like your prayers keep getting delivered to everyone else's address. What do we do when there is no guarantee of success, no promise of happily ever after, and no one shouting well done?

We show up anyway. Because at the end of the day, all we have in life is our own integrity. One day, I realized it really is not about reaching any particular destination as much as it's about living a life worthy of the journey. You can run to win, or you can run to learn. You can do the right thing even if no one ever sends a thank-you note. You can be intentional about who you are in this moment because this moment is all you have. You can live a life that is remembered by how you navigated what you were given rather than defined by what you got. Is it possible we can end up with something better than a happy ending someone else gave us? What if we finish our lives with a dignity we could only have given ourselves?

And that starts when *It's Not Your Turn* for accolades. In fact, there is no better place to start than when *It's Not Your Turn*.

> **Promise twelve to memorize:**
>
> "And though your beginning was small, your latter days will be very great." (Job 8:7 ESV)

DISCUSSION QUESTIONS

- ► What is the prayer you can pray—and ask someone else to pray—about your life right now?

- ► Do you believe prayer is effective? What experience makes that point?

- ► What's your next move?

Acknowledgments

T O MY HUSBAND, Seth, thank you for being the hand that has always grounded me. I will stand at the side of every stage you are standing on and be the loudest voice in the room. I love you, Chi.

To my kids, London, Hudson, and Sawyer, the greatest thing I'll ever do is be your mom.

To my parents, Joel and Vicki Thompson, my whole life, all I've ever wanted is to make you proud of me. Thank you for always putting us first.

To my sister and brother, Natasha Gordon and Joel Thompson Jr., you are my best friends. Saturdays on your couches are my safe spaces.

To my mother-in-law, Nicole Mattson, you have loved me like a daughter. I am blessed to call you mom.

To my best friend and co-blogger, Scarlett Longstreet, thanks for always being on the other end of my every call. I can't wait to watch you shine, Posey.

To my spiritual partner and friend, Vimbo Zvandasara, God knew what he was doing all those years ago. It's your turn, sis.

To my best friend since third grade, Jewel Jones, you are a once in a lifetime friend.

To Josue and Carla Vivanco, thank you for making Denver feel like we have family.

To my agent, Amanda Luedeke, you took a chance on me five years ago, and it gave me hope that kept me focused. Thank you for always being a sounding board.

To my editor, Al Hsu, this book is in the deepest debt to you. A good editor can make a good book, a great book. And that is what you did. Thank you.

To my publisher, InterVarsity Press, being able to be a part of a team that is committed to intellectual curiosity and faith is my deepest honor.

To all my sources in this book, without your research and insights this book wouldn't exist. Thank you for bringing so much to the conversation.

Notes

1. It's Not Your Turn

[1]Kristen Bialik and Richard Fry, "Millennial Life: How Young Adulthood Today Compares with Prior Generations," Pew Research Center Social and Demographic Trends, January 30, 2019, www.pewsocialtrends.org /essay/millennial-life-how-young-adulthood-today-compares-with-prior -generations/.

[2]Lulu Chang, "Facebook Use Associated with Decreased Health and Happiness, Study Finds," Digital Trends, May 26, 2017, www.digital trends.com/social-media/facebook-unhappy-unhealthy/.

[3]Bruce Y. Lee, "Depression Diagnoses Up 33% (Up 47% Among Millennials): Why There Is an Upside," *Forbes*, May 12, 2018, www.forbes.com /sites/brucelee/2018/05/12/depression-diagnoses-up-33-up-47-among -millennials-why-there-is-an-upside/#55db97415061.

[4]BluewaterTV, "Simon Sinek for Young People: Understand the Game," YouTube, September 23, 2018, www.youtube.com/watch?v=RfZGT _FcSLs, 30:40.

[5]Daniel Kahneman et al., "Would You Be Happier If You Were Richer? A Focusing Illusion," CEPS Working Paper No. 125, May 2006. www .princeton.edu/~ceps/workingpapers/125krueger.pdf.

[6]Richella Parham, *Mythical Me: Finding Freedom from Constant Comparison* (Downers Grove, IL: InterVarsity Press, 2019).

[7]Jodie Gummow, "7 Telltale Signs Social Media Is Killing Your Self-Esteem," *Salon*, March 11, 2014, www.salon.com/2014/03/11/7_telltale _signs_social_media_is_killing_your_self_esteem_partner/.

[8]Jodie Gummow, "Is Social Media Killing Your Self-Esteem?" Role Reboot, March 17, 2014, www.rolereboot.org/culture-and-politics/details/2014 -03-social-media-killing-self-esteem/.

2. It's Your Turn to Wait

[1]University of California Berkeley Haas School of Business, "Be More Patient? Imagine That: Neuroscientists Find Links Between Patience, Imagination in the Brain," *Science Daily*, April 4, 2017, www.sciencedaily.com/releases/2017/04/170404160028.htm.

[2]Ilene Strauss Cohen, "The Benefits of Delaying Gratification," *Psychology Today*, December 26, 2017, www.psychologytoday.com/us/blog/your-emotional-meter/201712/the-benefits-delaying-gratification.

[3]Cohen, "The Benefits of Delaying Gratification."

[4]Sarah A. Schnitker and Robert A. Emmons, "Patience as a Virtue: Religious and Psychological Perspectives," *Research in the Social Scientific Study of Religion* 18 (2007): 177-208, doi:10.1163/ej.9789004158511.i-301.69.

[5]Sarah A. Schnitker, "An Examination of Patience and Well-Being," *Journal of Positive Psychology* 7, no. 4 (June 26, 2012): 263-80, doi:10.1080/17439760.2012.697185.

[6]Schnitker and Emmons, "Patience as a Virtue."

3. It's Your Turn to Say It Out Loud

[1]Joseph A. DeVito, *The Interpersonal Communication Book* (Boston: Pearson Education, 2018).

[2]Kerry Patterson et al., *Crucial Conversations* (New York: McGraw-Hill Education, 2012), 5.

[3]Andrew B. Newberg and Mark Robert Waldman, *Words Can Change Your Brain* (New York: Avery, 2013), 24.

[4]Carol S. Dweck, *Mindset: The New Psychology of Success, How We Can Learn to Fulfill Our Potential* (New York: Ballantine Books, 2006), Introduction page.

[5]Newberg and Waldman, *Words Can Change Your Brain*, 24.

[6]Wendy L. Patrick, "The Way You Describe Others Is the Way People See You," *Psychology Today*, May 21, 2018, www.psychologytoday.com/us/blog/why-bad-looks-good/201805/the-way-you-describe-others-is-the-way-people-see-you.

[7]Newberg and Waldman, *Words Can Change Your Brain*, 24-25.

[8]*An Old Testament Commentary for English Readers, by Various Writers*, ed. Charles John Ellicott (London: Cassell, Petter & Galpin, 1882), https://biblehub.com/commentaries/ellicott/daniel/10.htm.

4. It's Your Turn to See

[1]Anna Almendrala, "'Indelible in the Hippocampus': Christine Blasey Ford Explains Science Behind Her Trauma," *HuffPost*, September 27, 2018, www.huffpost.com/entry/christine-blasey-ford-hippocampus-senate _n_5bad1a65e4b0b4d308d0edbc.

[2]Louise Delagran, "Impact of Fear and Anxiety," Taking Charge of Your Health & Wellbeing, 2016, www.takingcharge.csh.umn.edu/impact-fear -and-anxiety.

[3]Sue McGreevey, "Eight Weeks to a Better Brain," *Harvard Gazette*, September 12, 2019, https://news.harvard.edu/gazette/story/2011/01/eight -weeks-to-a-better-brain/.

[4]Emiliana R. Simon-Thomas, "Meditation Makes Us Act with Compassion," Greater Good Science Center, April 11, 2013, https://greatergood.berkeley .edu/article/item/meditation_causes_compassionate_action.

[5]Allen R. McConnell, "Gambling and the Self: A Sure Bet (To Lose Money)," *Psychology Today*, May 29, 2012, www.psychologytoday.com/us/blog/the -social-self/201205/gambling-and-the-self-sure-bet-lose-money.

5. It's Your Turn to Think Small

[1]Matthew Haag, "Thousands of Immigrant Children Said They Were Sexually Abused in U.S. Detention Centers, Report Says," *New York Times*, February 27, 2019, www.nytimes.com/2019/02/27/us/immigrant-children -sexual-abuse.html.

[2]Eric Barker, "The Simple Thing That Makes the Happiest People in the World So Happy," *Time*, April 13, 2014, https://time.com/59684/the-simple -thing-that-makes-the-happiest-people-in-the-world-so-happy/.

[3]Christopher Peterson, "Who Most Enjoy the Small Things in Life?" *Psychology Today*, June 1, 2010, www.psychologytoday.com/us/blog/the -good-life/201006/who-most-enjoy-the-small-things-in-life.

[4]Nana Dooreck, "6 Small Things the Best Leaders Do Differently," *Business Insider*, September 14, 2016, www.businessinsider.com/6-small-things -the-best-leaders-do-differently-2016-9.

[5]Jim Collins, *Good to Great* (New York: HarperCollins, 2001).

[6]Liz Wiseman, *Multipliers: How the Best Leaders Make Everyone Smarter* (New York: Harper Business, 2017), 55.

6. It's Your Turn to Set the Goal

[1]Keti Simmen-Janevska, Veronika Brandstätter, and Andreas Maercker, "The Overlooked Relationship Between Motivational Abilities and

Posttraumatic Stress: A Review," *European Journal of Psychotrauma-tology* 3, no. 1 (October 31, 2012), https://doi:10.3402/ejpt.v3i0.18560.

[2]David Burkus, "Are You Wasting Your 10,000 Hours?" *Forbes*, April 5, 2016, www.forbes.com/sites/davidburkus/2013/09/25/are-you-wasting -your-10000-hours/#31c70b8a303c.

[3]Mason Currey, *Daily Rituals: How Artists Work* (New York: Alfred A. Knopf, 2013).

[4]"The Science Behind Setting Goals (and Achieving Them)," *Forbes Books*, February 26, 2019, https://forbesbooks.com/the-science-behind-setting -goals-and-achieving-them/.

[5]Marilyn Price-Mitchell, "Goal-Setting Is Linked to Higher Achievement," *Psychology Today*, March 14, 2018, www.psychologytoday.com/us/blog /the-moment-youth/201803/goal-setting-is-linked-higher-achievement.

[6]Paul J. Meyer, *Attitude Is Everything* (Chicago: The Leading Edge Pub-lishing Company, 2006).

[7]Tom Corley, *Change Your Habits, Change Your Life: Strategies That Trans-formed 177 Average People into Self-Made Millionaires* (Minneapolis: North Loop Books, 2016).

[8]Mark Batterson, *The Circle Maker: Praying Circles Around Your Biggest Dreams and Greatest Fears*, exp. ed. (Grand Rapids, MI: Zondervan, 2016), 142.

[9]Thomas Oppong, "The Reading Brain (Why Your Brain Needs You to Read Every Day)," *Medium*, February 20, 2018, https://medium.com/@ alltopstartups/the-reading-brain-why-your-brain-needs-you-to-read -every-day-f5307c50d979.

[10]Dennis P. Carmody and Michael Lewis, "Brain Activation When Hearing One's Own and Others' Names," *Brain Research* 1116, no. 1 (October 20, 2006): 153-58, https://doi.org/10.1016/j.brainres.2006.07.121.

[11]W. Staffen et al., "Selective Brain Activity in Response to One's Own Name in the Persistent Vegetative State," *Journal of Neurology, Neuro-surgery & Psychiatry,* 77, no. 12 (2006): 1383-84, doi:10.1136/jnnp .2006.095166.

7. It's Your Turn to Network

[1]"Current World Population," Worldometers, accessed September 20, 2020, www.worldometers.info/world-population/.

[2]Shevali Best, "Mathematicians Reveal the Odds of Finding Love," *Daily Mail*, August 3, 2017, www.dailymail.co.uk/sciencetech/article-4757816 /Mathematicians-reveal-odds-finding-love.html.

[3]Malcolm Gladwell, *The Tipping Point: How Little Things Can Make a Big Difference* (Boston: Back Bay Books, 2013).

[4]Gladwell, *The Tipping Point.*

[5]Gladwell, *The Tipping Point,* 36.

[6]Gladwell, *The Tipping Point,* 37.

[7]Miriam Salpeter, "10 Tips to Becoming a Connector (Not Just a Networker)," *Vault,* August 3, 2016, www.vault.com/blogs/networking/10-tips -to-becoming-a-connector-not-just-a-networker.

[8]Lou Adler, "New Survey Reveals 85% of All Jobs Are Filled Via Networking," LinkedIn, February 29, 2016, www.linkedin.com/pulse/new -survey-reveals-85-all-jobs-filled-via-networking-lou-adler/.

[9]Falon Fetemi, "3 Habits of Highly Effective Networkers," *Forbes*, March 31, 2018, www.forbes.com/sites/falonfatemi/2018/03/31/three-habits -of-highly-effective-networkers/#6326aa815cf7.

[10]John Jecker and David Landy, "Liking a Person as a Function of Doing Him a Favour," *SAGE Human Relations* 22, no. 4 (April 22, 2016): 8, www.deepdyve .com/lp/sage/liking-a-person-as-a-function-of-doing-him-a-favour.

[11]Varsity Tutors, *The Autobiography of Benjamin Franklin Chapter Nine* (n.d.), www.varsitytutors.com/earlyamerica/lives-early-america/autobiography -benjamin-franklin/autobiography-benjamin-franklin-chapter-nine.

[12]Lindsay Holmes, "11 Ways Introverts Would Prefer to Start a Conversation," *HuffPost*, March 24, 2017, www.huffpost.com/entry/things -introverts-would-rather-chat-about-than-have-small-talk_n_58d167b 4e4b0be71dcf8962a.

[13]John Rougeux, "Why Only 10% of Your Facebook Followers See Your Posts," Causely, n.d., www.causely.com/blog/why-only-ten-percent -of-your-facebook-followers-see-your-posts.

8. It's Your Turn to Take a Second Look at Power

[1]Philip Zimbardo, *The Lucifer Effect: Understanding How Good People Turn Evil* (New York: Random House Trade Paperbacks, 2008).

[2]Zimbardo, *The Lucifer Effect,* 6.

[3]Nick Gillespie, "Millennials Are Selfish and Entitled, and Helicopter Parents Are to Blame," *Time*, August 21, 2014, https://time.com/3154186 /millennials-selfish-entitled-helicopter-parenting/.

[4]Zimbardo, *The Lucifer Effect,* 8.

[5]C. S. Lewis, *Mere Christianity* (New York: Simon and Schuster, 1943, 1980), 166.

[6]Lewis, *Mere Christianity,* 166.

[7]Jerry Useem, "Power Causes Brain Damage: How Leaders Lose Mental Capacities—Most Notably for Reading Other People—That Were Essential to Their Rise," *Atlantic,* June 23, 2017, www.theatlantic.com /magazine/archive/2017/07/power-causes-brain-damage/528711/.

[8]Useem, "Power Causes Brain Damage."

[9]Adam D. Galinsky et al., "Losing Touch: Power Diminishes Perception and Perspective," Kellogg School of Management at Northwestern University, November 1, 2009, https://insight.kellogg.northwestern.edu/article /losing-touch.

[10]Sonja K. Foss, *Rhetorical Criticism: Exploration and Practice,* 5th ed. (Long Grove, IL: Waveland Press, 2017).

[11]Foss, *Rhetorical Criticism,* 239.

[12]N. T. Feather, "Acceptance and Rejection of Arguments in Relation to Attitude Strength, Critical Ability, and Intolerance of Inconsistency," *Journal of Abnormal and Social Psychology* 69, no. 2 (1964): 127-36.

[13]Feather, "Acceptance and Rejection of Arguments," 138.

[14]Mary E. Wheeler and Susan T. Fiske, "Controlling Racial Prejudice: Social-Cognitive Goals Affect Amygdala and Stereotype Activation," *Psychological Science* 16, no. 1 (January 1, 2005): 56-63.

[15]"Ahmaud Arbery: What Do We Know About the Case?" BBC News, June 5, 2020, www.bbc.com/news/world-us-canada-52623151.

[16]Jamiles Lartey, "Why It's Not So Simple to Arrest the Cops Who Shot Breonna Taylor," The Marshall Project, August 8, 2020, www.themarshall project.org/2020/08/08/why-it-s-not-so-simple-to-arrest-the-cops-who -shot-breonna-taylor.

9. It's Your Turn to Find Community

[1]Richard J. Krejcir, "Statistics on Pastors: 2016 Update Research on the Happenings in Pastors' Personal And Church Lives," Francis A. Schaeffer Institute of Church Leadership Development, 2016, https://files.stablerack .com/webfiles/71795/pastorsstatWP2016.pdf.

[2]"Hate Crimes," FBI.gov, accessed May 3, 2016, www.fbi.gov/investigate /civil-rights/hate-crimes.

[3]Colleen M. Cowgill, Kimberly Rios, and Ain Simpson, "Generous Heathens? Reputational Concerns and Atheists' Behavior Toward Christians in Economic Games," *Science Direct,* July 10, 2017, www.sciencedirect .com/science/article/abs/pii/S0022103116307910.

[4]"Americans Divided on the Importance of Church," Barna Group, March 24, 2014, www.barna.com/research/americans-divided-on-the-importance-of -church/.

[5]Carol Zimmermann, "More Americans Believe in Higher Power Than in God, Study Says," Crux, April 27, 2018, https://cruxnow.com/church-in -the-usa/2018/04/more-americans-believe-in-higher-power-than-in-god -study-says/.

[6]Caryle Murphy, "Most Americans Believe in Heaven . . . and Hell," Pew Research Center, August 17, 2020, www.pewresearch.org/fact-tank /2015/11/10/most-americans-believe-in-heaven-and-hell/.

[7]W. Gardner Selby, "Texan Says U.S. Muslims Lately Subject to More Attacks, Hate Crimes Than Ever," *Politifact*, August 3, 2018, www.politifact .com/factchecks/2018/aug/03/omar-rachid/US-Muslims-more-attacks -hate-crimes-than-ever-be/.

[8]Beyza Binnur Donmez, "White Extremism Under Trump Presidency Rises 55%: Study," Anadolu Agency, March 19, 2020, www.aa.com.tr /en/americas/white-extremism-under-trump-presidency-rises-55-study /1772045.

[9]Andrew Newberg and Mark Waldman, *How God Changes Your Brain: Breakthrough Findings from a Leading Neuroscientist* (New York: Ballantine Books, 2010), 29.

[10]Sonja K. Foss, *Rhetorical Criticism: Exploration and Practice*, 5th ed. (Long Grove, IL: Waveland Press, 2017).

[11]Joey Marshall, "Are Religious People Happier, Healthier? Our New Global Study Explores This Question," Pew Research Center, January 13, 2019, www.pewresearch.org/fact-tank/2019/01/31/are-religious-people -happier-healthier-our-new-global-study-explores-this-question/.

[12]Newberg and Waldman, *How God Changes Your Brain*, 60.

[13]David Buettner, "These Traditional Diets Can Lead to Long Lives," *National Geographic*, January 2, 2020, www.nationalgeographic.com/magazine /2020/01/these-traditional-diets-from-the-blue-zones-can-lead -to-long-lives-feature/.

[14]Kenneth I. Pargament et al., "Religious Struggle as a Predictor of Mortality Among Medically Ill Elderly Patients: 2-Year Longitudinal Study," *Archives of Internal Medicine* 161, no.15 (August 13, 2001): 1881-5.

[15]Jenny Santi and Deepak Chopra, *The Giving Way to Happiness: Stories and Science Behind the Life-Changing Power of Giving* (New York: TarcherPerigee, 2016).

[16]Jenny Santi, "The Secret to Happiness Is Helping Others," *Time*, n.d., https://time.com/collection/guide-to-happiness/4070299/secret-to-happiness/.

[17]Annamarie Mann, "Why We Need Best Friends at Work," Gallup, January 15, 2018, www.gallup.com/workplace/236213/why-need-best-friends-work.aspx.

[18]Brené Brown, "The Power of Vulnerability," filmed at TEDxHouston, Houston, TX, video, 20:49, www.ted.com/talks/brene_brown_the_power_of_vulnerability/details?language=en.

[19]Tom Rath and Jim Harter, "Your Friends and Your Social Well-Being," Gallup, August 19, 2010, https://news.gallup.com/businessjournal/127043/friends-social-wellbeing.aspx.

[20]Joseph A. DeVito, *The Interpersonal Communication Book* (Boston: Pearson Education, 2018).

[21]Jonathan Morrow, "Only 4 Percent of Gen Z Have a Biblical Worldview," Impact 360 Institute, n.d., www.impact360institute.org/articles/4-percent-gen-z-biblical-worldview/.

[22]"Americans Divided on the Importance of Church," Barna Group, March 24, 2014, www.barna.com/research/americans-divided-on-the-importance-of-church/#.V-hxhLVy6FD.

[23]Kevin Donnelly, "Marketing to Millennials: 5 Massive Trends That Are Leading the Way," Shopify, February 10, 2016, www.shopify.com/blog/75614533-marketing-to-millennials-5-massive-trends-that-are-leading-the-way.

[24]Liz Mineo, "Good Genes Are Nice, but Joy Is Better: Over Nearly 80 Years, Harvard Study Has Been Showing How to Live a Healthy and Happy Life," *Harvard Gazette*, April 11, 2017, https://news.harvard.edu/gazette/story/2017/04/over-nearly-80-years-harvard-study-has-been-showing-how-to-live-a-healthy-and-happy-life/.

10. It's Your Turn to Re-envision God

[1]Chunliang Feng et al., "Social Hierarchy Modulates Neural Responses of Empathy for Pain," *Social Cognitive and Affective Neuroscience* 11, no. 3 (March 2016): 485-95, doi:10.1093/scan/nsv135.

[2]Heidi Grant Halvorson, "The Surprising Reason Why We Break Promises," *HuffPost*, March 13, 2013, www.huffpost.com/entry/breaking-promises_b_2449631.

[3]Peter M. Gollwitzer et al., "When Intentions Go Public," *Psychological Science* 20, no. 5 (May 1, 2009): 612-18, doi:10.1111/j.1467-9280.2009.02336.x.

[4]"Time with God: An Interview with J. I. Packer," C. S. Lewis Institute, Winter, 2016, www.cslewisinstitute.org/Time_with_God_An_Interview _with_JI_Packer_FullArticle.

[5]Andrew Newberg and Mark Waldman, *How God Changes Your Brain: Breakthrough Findings from a Leading Neuroscientist* (New York: Ballantine Books, 2010), 110.

[6]Newberg and Waldman, *How God Changes Your Brain*, 11.

[7]Thomas Armstrong, "The Stages of Faith According to James W. Fowler," American Institute for Learning and Human Development, June 12, 2020, www.institute4learning.com/2020/06/12/the-stages-of-faith-according -to-james-w-fowler/.

[8]Newberg and Waldman, *How God Changes Your Brain*, 11.

[9]Newberg and Waldman, *How God Changes Your Brain*, 14.

[10]Newberg and Waldman, *How God Changes Your Brain,* 14.

[11]"American Piety in the 21st Century: New Insights to the Depth and Complexity of Religion in the U.S.," Baylor University, September 2006, www.baylor.edu/content/services/document.php/33304.pdf.

[12]Robert H. Gass and John S. Seiter, *Persuasion: Social Influence and Compliance Gaining*, 6th ed. (New York: Routledge, 2018).

[13]Bob Altemeyer, "To Thine Own Self Be Untrue: Self-awareness and Authoritarians," *North American Journal of Psychology* 1, no. 2 (1999): 157-64.

11. It's Your Turn to Move on Maybe

[1]"A Class Divided (full film)," Frontline PBS, January 18, 2019, YouTube, https://www.youtube.com/watch?v=1mcCLm_LwpE, 53:00.

[2]"Risk-Takers Are Smarter, According to a New Study," Science Daily, November 30, 2015, www.sciencedaily.com/releases/2015/11/151130113545 .htm.

[3]Jennifer Warner, "Are Risk Takers Happier?" WebMD, September 19, 2005, www.webmd.com/balance/news/20050919/are-risk-takers-happier.

[4]Jackie Swift, "How We Make Decisions and Take Risks," Cornell Research, n.d., https://research.cornell.edu/news-features/how-we-make -decisions-and-take-risks.

[5]Sam McRoberts, "Here's What Science Says You Should Do to Achieve Greater Success," *Entrepreneur*, December 29, 2017, www.entrepreneur .com/article/305985.

[6]Frances Bridges, "5 Ways to Be Brave According to Brené Brown's Netflix Special 'The Call to Courage,'" *Forbes*, April 29, 2019, www

.forbes.com/sites/francesbridges/2019/04/29/5-ways-to-be-brave
-according-to-brene-browns-netflix-special-the-call-to-courage
/#52b09a4179f4.

[7]Bridges, "5 Ways to Be Brave."

[8]David Topus, *Talk to Strangers: How Everyday Random Encounters Can Expand Your Business, Career, Income, and Life* (Hoboken, NJ: Wiley, 2012).

[9]Megan Sagar, "A Very Loud Reminder of All the Songs Taylor Swift Has Solo-Written," UMusic, n.d., https://umusic.co.nz/umusic/pop/a-very-loud-reminder-of-all-the-songs-taylor-swift-has-solo-written/.

[10]Mel Robbins, "The Five Elements of the 5 Second Rule," Mel Robbins, April 25, 2018, https://melrobbins.com/five-elements-5-second-rule/.

[11]J. D. Reed, "Mind Over Matter," *New York Times*, March 9, 2003, www.nytimes.com/2003/03/09/nyregion/mind-over-matter.html.

[12]Dean Radin et al., "Compassionate Intention as a Therapeutic Intervention by Partners of Cancer Patients: Effects of Distant Intention on the Patients Autonomic Nervous System," *Science Direct* 4, no. 4 (July 2008): 235-43, https://doi.org/10.1016/j.explore.2008.04.002.

12. It's Your Turn to Make Your Move

[1]"Pastor Wintley Phipps Predicted Oprah's Success," Oprah Winfrey Network, November 22, 2015, YouTube, www.youtube.com/watch?v=-gEpwe7gktM, 1:25.

[2]Scarlett Longstreet (@Scarlett Posner), correspondence with the author, September 13, 2020.

[3]Seth Day (@SethMDay), correspondence with the author, September 20, 2020.

[4]Vimbo Zvandasara (@Ohvimbo), correspondence with the author, September 16, 2020.

[5]Carla Hernandez (@CarlaLilianaV), correspondence with the author, September 6, 2020.

[6]Alex Gordon, correspondence with the author, September 13, 2020.

[7]Flor Osorio Just (@lipsnstockings), correspondence with the author, September 16, 2020.

[8]Mary Beth Minnis (@MaryBethMinnis), correspondence with the author, October 4, 2020.

About the Author

Dr. Heather Thompson Day is an interdenominational speaker and contributor for Religion News Service, *Newsweek,* and the Barna Group. She is associate professor of communication and rhetoric at Colorado Christian University. She is passionate about supporting women and runs an online community called *I'm That Wife,* which has nearly two hundred thousand followers.

Heather's writing has been featured on the *TODAY Show* and the National Communication Association. She has been interviewed by *BBC Radio Live*. She believes her calling is to stand in the gaps of our churches for young people. She is the author of seven books, including *Confessions of a Christian Wife*.

She resides in Denver, Colorado, with her husband, Pastor Seth Day, and their three children, London, Hudson, and Sawyer Day.

You can contact Heather at:
www.HeatherThompsonDay.com
Twitter: @HeatherTDay
Instagram and Facebook: @HeatherThompsonDay
Blog on all social channels: @ImThatWife